D0844382

WHAT EVERY ENGINEER SHOULD KNOW ABOUT

PROJECT MANAGEMENT

WHAT EVERY ENGINEER SHOULD KNOW

A Series

Editor

William H. Middendorf

Department of Electrical and Computer Engineering
University of Cincinnati
Cincinnati, Ohio

1. What Every Engineer Should Know About Patents, *William G. Konold, Bruce Tittel, Donald F. Frei, and David S. Stallard*
2. What Every Engineer Should Know About Product Liability, *James F. Thorpe and William H. Middendorf*
3. What Every Engineer Should Know About Microcomputers: Hardware/Software Design, A Step-by-Step Example, *William S. Bennett and Carl F. Evert, Jr.*
4. What Every Engineer Should Know About Economic Decision Analysis, *Dean S. Shupe*
5. What Every Engineer Should Know About Human Resources Management, *Desmond D. Martin and Richard L. Shell*
6. What Every Engineer Should Know About Manufacturing Cost Estimating, *Eric M. Malstrom*
7. What Every Engineer Should Know About Inventing, *William H. Middendorf*
8. What Every Engineer Should Know About Technology Transfer and Innovation, *Louis N. Mogavero and Robert S. Shane*
9. What Every Engineer Should Know About Project Management, *Arnold M. Ruskin and W. Eugene Estes*
10. What Every Engineer Should Know About Computer-Aided Design and Computer-Aided Manufacturing: The CAD/CAM Revolution, *John K. Krouse*
11. What Every Engineer Should Know About Robots, *Maurice I. Zeldman*
12. What Every Engineer Should Know About Microcomputer Systems Design and Debugging, *Bill Wray and Bill Crawford*
13. What Every Engineer Should Know About Engineering Information Resources, *Margaret T. Schenk and James K. Webster*
14. What Every Engineer Should Know About Microcomputer Program Design, *Keith R. Wehmeyer*
15. What Every Engineer Should Know About Computer Modeling and Simulation, *Don M. Ingels*

16. What Every Engineer Should Know About Engineering Workstations, *Justin E. Harlow III*
17. What Every Engineer Should Know About Practical CAD/CAM Applications, *John Stark*
18. What Every Engineer Should Know About Threaded Fasteners: Materials and Design, *Alexander Blake*
19. What Every Engineer Should Know About Data Communications, *Carl Stephen Clifton*
20. What Every Engineer Should Know About Material and Component Failure, Failure Analysis, and Litigation, *Lawrence E. Murr*
21. What Every Engineer Should Know About Corrosion, *Philip Schweitzer*
22. What Every Engineer Should Know About Lasers, *D. C. Winburn*
23. What Every Engineer Should Know About Finite Element Analysis, *edited by John R. Brauer*
24. What Every Engineer Should Know About Patents: Second Edition, *William G. Konold, Bruce Tittel, Donald F. Frei, and David S. Stallard*
25. What Every Engineer Should Know About Electronic Communications Systems, *L. R. McKay*
26. What Every Engineer Should Know About Quality Control, *Thomas Pyzdek*
27. What Every Engineer Should Know About Microcomputers: Hardware/Software Design, A Step-by-Step Example. Second Edition, Revised and Expanded, *William S. Bennett, Carl F. Evert, and Leslie C. Lander*
28. What Every Engineer Should Know About Ceramics, *Solomon Musikant*
29. What Every Engineer Should Know About Developing Plastics Products, *Bruce C. Wendle*
30. What Every Engineer Should Know About Reliability and Risk Analysis, *M. Modarres*
31. What Every Engineer Should Know About Finite Element Analysis: Second Edition, Revised and Expanded, *edited by John R. Brauer*
32. What Every Engineer Should Know About Accounting and Finance, *Jae K. Shim and Norman Henteleff*
33. What Every Engineer Should Know About Project Management: Second Edition, Revised and Expanded, *Arnold M. Ruskin and W. Eugene Estes*

ADDITIONAL VOLUMES IN PREPARATION

WHAT EVERY ENGINEER SHOULD KNOW ABOUT

PROJECT MANAGEMENT

Second Edition, Revised and Expanded

Arnold M. Ruskin

Partner

Claremont Consulting Group

La Cañada, California

W. Eugene Estes

Consulting Civil Engineer

Westlake Village, California

CRC Press is an imprint of the
Taylor & Francis Group, an **informa** business

Library of Congress Cataloging-in-Publication Data

Ruskin, Arnold M.
What every engineer should know about project management / Arnold M. Ruskin, W. Eugene Estes. -- 2nd Ed., rev. and expanded.
pm cm. -- (What every engineer should know ; 33)
Includes bibliographical references and index.
ISBN 0-8247-8953-9 (acid-free paper)
1. Engineering--Management. I. Estes, W. Eugene
II. Title. III. Series
TA190.R87 1994 94-39689
658.4'04--dc20 CIP

The publisher offers discounts on this book when ordered in bulk quantities. For more information, write to Special Sales/Professional Marketing at the address below.

CRC Press
6000 Broken Sound Parkway, NW
Suite 300, Boca Raton, FL 33487

270 Madison Avenue
New York, NY 10016

2 Park Square, Milton Park
Abingdon, Oxon OX14 4RN, UK

Preface to the Second Edition

Project management has evolved at a pace that we could hardly imagine when we wrote the first edition. Both software engineers and system engineers have influenced project management considerably as they have brought their particular insights and approaches to projects. Prominent among their contributions are improvements in requirements specification, work breakdown structures, project control, and risk management. This second edition was motivated primarily by the need to describe these advances to our readers.

Having decided to prepare a second edition, we also took the opportunity to fine-tune the rest of the book. We have adjusted the discussions of the anatomy of a project, the roles and responsibilities of the project manager, and negotiating, and we have added new material on motivation, matrix arrangements, and project records. While matrix arrangements are not yet ubiquitous, they are becoming so common that every engineer should know how to operate in a matrix situation.

This book is written by engineers for engineers, but it is not for engineers alone. Many nonengineers read the first edition because they happened to attend a seminar where the book was used. We have been gratified by their comments that the book serves them well. We hope that both our engineering and nonengineering audiences find this second edition even more useful.

Arnold M. Ruskin
W. Eugene Estes

Preface to the First Edition

This book presents basic concepts and tools of projects and project management. Projects date from the earliest days of civilization and include the building of the Egyptian pyramids and the Roman roads and aqueducts. Today, projects are organized not only for building great public works, but also for such diverse tasks as performing applied research, developing software, installing equipment, building extensive and complex systems, shutting down and renovating major facilities, and preparing proposals and studies. Few, if any, engineers can totally escape project work in the 1980s.

Our goals in this book are to provide a fundamental and broad view of project activity that applies to all engineers and to introduce some useful tools to the novice project manager. We have organized the book around the duties of the project manager because project management is what makes project work different from other technical work. The topics are important, however, to all who are associated with projects, whether they are project managers, supervisors of project managers, managers of subprojects and tasks, project staff members, others who support projects, or project customers. Likewise, the topics apply to engineers in all kinds of settings, including industrial firms, governmental agencies, consulting firms, and literally all organizations where particular objectives must be accomplished and accomplished on time and within budget.

For simplicity, we refer to the person or group who is served by a project as the customer. The customer may in fact be a client of the project manager's organization, but may just as likely be the project manager's boss, the organization's chief engineer, its marketing vice president, or some other individual or group within the organization. Whatever the case, the same principles apply.

All successful project work rests on a foundation of setting objectives, planning, directing and coordinating, controlling, reporting, and negotiating. We discuss all of these. Certain aspects of these topics are not limited to project work and might be considered by some to lie outside the realm of a book on project management. We include them because we want the reader to know how broad a project manager's duties are. People fall down in project management when they do not address their full range of duties and their consequences.

There is little that is truly mysterious about managing projects. It may only seem so because of the many facets involved. Our aim is to identify these facets and make them clear so that any mystery is removed. We hope that our reader will obtain the awareness, understanding, and basic equipment needed first to approach a project assignment with a healthy mix of respect and confidence and then to complete it successfully.

Arnold M. Ruskin
W. Eugene Estes

Contents

Figures and Tables

FIGURES

TABLES

Introduction

Project management can be learned. While some people seem to know intuitively how to manage projects, most effective project managers learn their skills. This book explains how to manage projects so that their desired results are obtained on schedule and within budget.

Skillful project management involves knowing what is to be done, who is to do it, and when and how it should be done. This book considers all these factors and also explains why. After all, there is little in the affairs of men and women that does not require judgment. If project managers are to apply their judgment wisely, they need to know the rationale for each task as well as the task itself.

The skills described in this book apply to projects large and small, to design efforts and construction jobs, to field studies, laboratory investigations, and software development. Indeed they apply to every kind of project. True, large or complicated projects offer more opportunities to overlook critical aspects than small or simple projects, but small and simple projects may be so stringently scheduled or budgeted that they provide no margin for error. In either case, therefore, project managers must pay attention to the factors discussed in the following chapters, whether explicitly or implicitly, if they are to ensure successful projects.

As a text on project management in general, this book makes no attempt to address the nuances peculiar to selected types of projects. Rather, we assume that our readers will supply their own insights into their particular situations and thus season our basic principles to suit their own tastes.

As noted, this book is a mixture of the what, who, when, how, and why of project management. Some background, however, is given first so that the context of these topics is clear. For this reason, Chapter 1 describes and characterizes projects. We then move in the following chapter to the roles and responsibilities of the project manager; in this chapter we begin to introduce some tools that the project manager can use and outline the functions of others that are substantial enough to warrant their own chapters. Beginning in Chapter 3 and continuing to Chapter 9, we take up individual tools and techniques that most project managers find occasion to use from time to time. Finally, in the epilogue we try to show the interdependence of the tools and techniques when they are used to manage projects.

Everyone manages projects, but some manage better than others. In many cases, the consequences of inept management are not terribly significant. If it takes an extra weekend to build the children's playhouse, no one will really suffer. But if it takes an additional three months and $2 million to complete a passenger terminal, the consequences are serious. While the playhouse construction manager can afford to be inept, the terminal construction manager does not have this luxury. Nor does any engineer who manages resources on behalf of someone else.

1

Anatomy of a Project

A project is a special kind of activity. It involves something that is both unique and important and thereby requires unusual attention. It also has boundaries with other activities so that its extent is defined. And it has a beginning and an end and objectives whose accomplishment signals the end.

A project is different from never-ending functions, such as the accounting function, the manufacturing function, the sales function, the personnel function, and so forth. Note, however, that these functions may contain projects within them. Projects are also different from activities that have beginnings and ends but no specific goal, such as having dinner, playing the violin, and watching television. And projects differ from programs, whose conclusion is diffuse, and from other activities that have no bounds, such as the sum of all the affairs of a major corporation or a governmental agency. Here too, though, there may be projects within the program or within the total range of affairs.

I. PROJECT LIFE CYCLES

Because a project has a beginning and an end, it has a life cycle. The life cycle starts with a concept phase and concludes with a postaccomplishment phase, as shown in Figure 1.1. Four intermediate phases are also identified in the figure: a project definition or proposal preparation phase, a planning and organizing phase, a plan validation phase, and a performance or work accomplishment phase. Each succeeding phase is more concrete than the preceding one, as the project matures from an overall concept to a set of tasks that in their totality accomplish the project.

A. Concept Phase

The concept phase begins with the initial notion or "gleam in the eye" of someone who imagines accomplishing some objective. The objective may be to provide a bridge, to develop a manufacturing capability, to obtain information, to make arrangements to accomplish some task or goal, to build relationships to enable future actions, to win some number of customers, to train some number of employees, and so forth. Sometimes the objective is less specific than, say, to provide a bridge. Perhaps the objective is stated merely as "to provide a means for 1,500 cars to cross the river per hour." This approach is just as valid as specifying a bridge, and it may be more useful because it does not rule out choices that may be better.

The only absolute restriction on defining an objective is that one must be able to tell when it has been attained. Generally, however, the objectives statement can be usefully rephrased as a succinct set of individual requirements, which will make the project more manageable. Each requirement in a well-formulated set has the following characteristics, as appropriate:

1. Defines *what,* but not *how,* to the maximum extent possible, to allow the project team to select the most appropriate approach

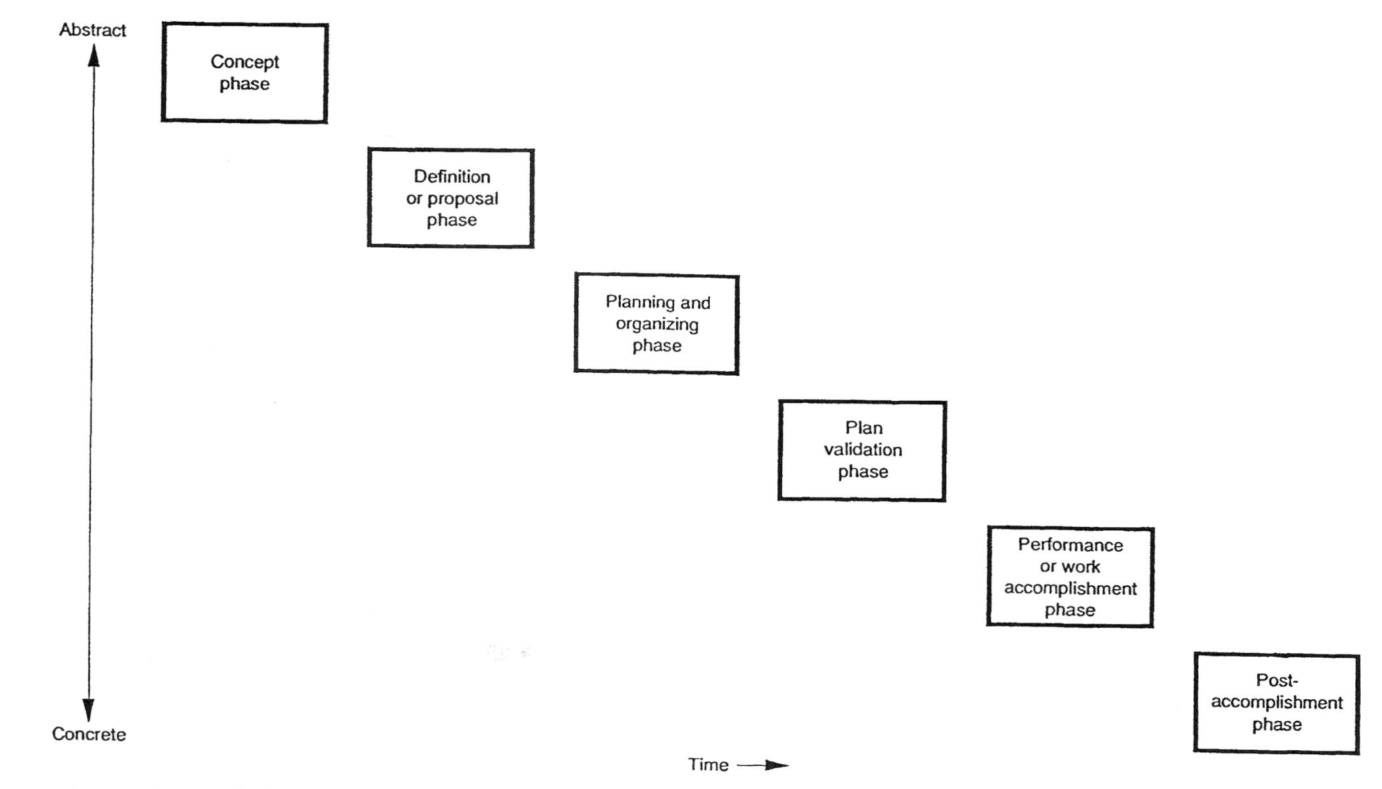

Figure 1.1. Idealized phases of a project.

2. Is expressed as "shall" or "shall not," to enable quick recognition of a requirement*
3. Is simple or "atomic," not compound, to provide clarity and to facilitate change control
4. Can be validated by testing, demonstration, analysis, or inspection to a precision that is as good as or better than the requirement's precision, to determine when the requirement has been met
5. Identifies whether the situation to which it applies is of nominal (most likely), worst case, or expected values, to guide those whose products must satisfy the requirement
6. Identifies when the requirement shall first be met and when it shall no longer be needed
7. Identifies the conditions or mode to which it applies, e.g., always or sometimes (when), also to guide those whose products must satisfy the requirement
8. Identifies related requirements, to help evaluate the impacts of proposed changes
9. Identifies its source, including whether it serves end-users or just intermediate developments, to indicate who must be consulted if changes to it are contemplated
10. Can be satisfied by its due date with available or attainable resources
11. Has a unique label, to enable its bookkeeping
12. Has a custodian, to localize information about it and its implementation status

---

*If only "shall" statements are used, a wordprocessor can easily find all the requirements statements. While other forms, e.g., "must," are grammatically equivalent, their use opens the door to an unlimited set of alternatives and makes identification of all requirements statements uncertain.

"Should" statements indicate desires, not requirements. They are acceptable, for often they can be accommodated without penalty. Nevertheless, their fulfillment is optional with the project team and they are not the basis for contractual obligations.

As indicated in item 1, the concept phase does *not* include how to accomplish the objectives—such considerations duly come later. Many projects experience grave difficulties because the concept phase is truncated before it is finished and attention is prematurely turned toward the means for accomplishing the objective. A project's objectives need to be fully explored and developed in the concept phase if their means for accomplishment are to be optimized. Otherwise, much time will be spent in needless arguments about the approach because different people will have different ideas about what is to be accomplished.

The concept phase not only includes formulation of project objectives but also identification of relevant constraints. This is not to say that the list of constraints must be exhaustive and that further constraints will not be identified. Indeed, it is a rare project where such foresight exists. Sometimes, however, an objective must necessarily be accomplished within certain limits, such as budget or time, or must be accomplished using certain tools, personnel, or procedures. In these cases, the constraint is so important that it needs to be stated along with the objectives. Otherwise, the project could conceivably be developed in a way that violates a cardinal limitation.

Formulating objectives fully during the concept phase is a major help toward an efficient and relatively peaceful project. It does not guarantee, however, that they will not later be reviewed or reconsidered. As the project matures, the customer* may change its objectives, or the project's activities may produce information showing that the objectives

---

*Depending upon the situation, *customer* may mean (1) the client's representative, (2) the client's own customer, (3) a group of people in the client's organization, (4) the project manager's boss, (5) another individual in the project manager's own organization, or (6) a group of people in the project manager's organization.

are not fully appropriate. In either case, the project should return to the concept phase to confirm or change the objectives. If the objectives are changed, everything else that follows is also subject to being changed!

When the concept is developed to the point that it can be meaningfully discussed and it can be concluded that it has a reasonable chance of solving or resolving the problem at hand, the project is ready to advance to the next stage, project definition.

B. Project Definition Phase

Project definition (and/or proposal preparation) follows the concept phase. This phase has two parts. The first comprises characterizing the project in terms of assumptions about the situation, alternative ways of achieving the objectives, decision criteria and models for choosing among viable alternatives, practical constraints, significant potential obstacles, and resource budgets and schedules needed to implement the viable alternatives. The second part consists of tentatively selecting the overall approach that will be used to achieve the objectives. Obviously, not everything that eventually needs to be known in order to accomplish the project is known at this early stage. Thus, many choices are made tentatively, with contingency arrangements identified in case the choices are found to be unsatisfactory.

If the amount of uncertainty is so great that the contingency allowances are unacceptably high, two remedies are available. First, the project can be divided into two sequential subprojects. The objective of the first subproject is to obtain information that will reduce the amount of uncertainty. This information is then incorporated in a second subproject directed toward achieving the main objective. Second, one can merely pursue the several alternatives in parallel until it becomes clear which to continue and which to abandon. This second approach will generally be more expensive than the first, but it *may* take less elapsed time to reach the ultimate objective.

The project definition produces a plainly written, unambiguous description of the project in sufficient detail to support a proposal or request to the customer to undertake the overall project. The definition should address

1. How the work will be done
2. How the project will be organized
3. Who the key personnel are
4. A tentative schedule
5. A tentative budget

The aim is to convince the customer that the doers know what to do and are qualified to do it. At the same time, the description should not have so much detail that the project is essentially planned or completed even before it is authorized.* Thus, the project definition phase represents a beginning-to-end thinking-through of the project but does not accomplish the project in and of itself. It is like a map of a route rather than the route per se, and it is a coarse map at that.

If at all possible, key personnel of the would-be project should be involved in defining the project and preparing the proposal. Sometimes this ideal arrangement is not possible for legitimate reasons such as unavailability at the moment. When key personnel are not significantly involved at this stage, it is often worth the expense of money and time for them to review the work of this phase and perhaps modify

*One always has the problem in preparing project proposals of deciding how much of the actual project to do in order to show that it can in fact be done. In general, the answer is that as much should be done as necessary to obtain project authorization, but no more. The exact amount is a matter of judgment and depends upon a variety of factors such as the novelty of the project, the customer's familiarity with the subject and with the project team, and the competitive situation.

it or convert it to something they can support before progressing to the planning phase.

C. Planning and Organizing Phase

Assuming that the customer accepts the project proposal and authorizes the project, the next phase is to plan and organize it. It is in this phase that *detailed* plans are prepared and tasks identified, with appropriate milestones, budgets, and resource requirements established for each task.

Some project managers try to do this work during the project definition phase, partly to demonstrate that they know how to manage the project, and then they skip lightly over it once the project is authorized. While the desire to economize is commendable, it is not always effective. Indeed, such early efforts are seldom sufficient to actually manage the project, and the overall costs are invariably higher than if proper planning is done. Moreover, the project's scope of work may be revised between the time it is proposed and the time it is authorized, thereby invalidating some parts of a preliminary plan.

This phase also includes building the organization that will execute the project. While some consideration is undoubtedly given during the project definition phase to how the project will be staffed, there is typically no guarantee at that early time that the project will actually be authorized and that the individuals can be committed to the effort. Furthermore, plans are most effective when they are developed by the doers, and doers are best motivated by having a say in planning their work. Therefore, it behooves project managers to develop their organizations and their plans simultaneously, in order to have each enhance the other.

To organize the project team, the project manager must identify the nature, number, and timing of the different skills and traits needed and arrange for them to be available as required. These requirements include not only various sorts of technical expertise, but also skills and traits in such areas as communication, leadership, "followership," conceptualization, analysis, detail-following, initiative, resourcefulness, wisdom, enthusiasm, tolerance for ambiguity, need for specific information, and so forth. No single prescription can be given for all projects in terms of what is required. Suffice it to say that there should be an appropriate range and mix of the characteristics needed for the project to succeed.

Much more is said about project planning in Chapter 3.

D. Plan Validation Phase

Once the project has been planned and organized, it is time to begin doing it. Haste makes waste, however, and most project managers will help themselves considerably by deliberately validating their plans before executing them. Plan validation consists of literature searches, field reconnaissances, experiments, interviews, and other forms of gathering data and information that (1) validate or rectify any assumptions made in the plan and (2) identify and characterize critical aspects of the project so that it can go forward smoothly.

Flaws in assumptions that are revealed during plan validation will show up later if the validation is not performed. If they are discovered late, however, valuable time and possibly resources will have been squandered. Moreover, the remaining time and resources may be inadequate to achieve the objectives.

Even if assumptions are validated at this time, some of them may nevertheless turn out later to be faulty. Thus, not only should assumptions be examined during this phase for flaws, but adverse consequences that could arise from assumptions with hidden flaws should also be considered. Dealing with this latter potentiality is called risk planning or risk management and is the subject of Chapter 5.

Insight, candor, and a realistic outlook on the part of the project team are necessary to validate plans properly. Insight is needed to identify the project's areas of vulnerability in order to identify where validation and risk planning are most useful. Candor is needed to be able to voice these concerns. And a realistic outlook is needed to keep from dismissing them as insignificant. If any one of these ingredients is missing, the project has a poor chance of success. Project managers do not need such a burden, so it behooves them to pay proper attention to plan validation and risk management.

If validation and risk-planning activities confirm the team's fears about being successful, the team should revisit the concept, definition, and planning phases in consultation with its customer. If validation indicates, however, that success is within reach, then it is time to proceed to the performance phase.

E. Performance Phase

The performance, or work accomplishment, phase is the part of the project that most people think of when they think of a project. Basically it consists of doing the work and reporting the results. Doing the work includes directing and coordinating other people and controlling their accomplishments so that their collective efforts achieve the project's objectives. These topics warrant entire chapters. Directing and coordinating work is discussed in Chapters 6 and 7, and controlling work is discussed in Chapter 4.

F. Post-Accomplishment Phase

It is commonly believed that completing the last task in the performance or work accomplishment phase concludes the project, but there is yet a final phase called a post-accomplishment phase. This phase consists of

1. Confirming that the customer is satisfied with the work and performing any small adjustments and answering any questions necessary to achieve satisfaction

2. Putting project files in good order so that they will be useful for future reference
3. Restoring equipment and facilities to appropriate status for later use or decommissioning them
4. Assuring that project accounts are brought up-to-date, appropriately audited, and closed out
5. Assisting project staff in being reassigned
6. Paying any outstanding charges
7. Collecting any fees or payments due

G. Phase-to-Phase Relationships

While Figure 1.1 and the preceding discussion tend to make the interphase boundaries distinct, the successive phases often overlap in time. As the project is defined, for example, second thoughts about the concept may arise, leading to its revision. Thus, the concept phase is extended so that it overlaps the project definition phase. Such overlaps are shown in Figure 1.2.

In addition to phases overlapping, it is likely that some iteration among, say, the plan validation phase, the concept phase, and the project definition and planning phases will occur. Certainly there are cases where the results of plan validation have caused the entire project to be revamped.

II. PROJECT BOUNDARIES, INPUTS, OUTPUTS, AND INTERFACES

A. Project Boundaries

Project managers should recognize the boundaries of their domains. Elements that lie within a project's boundaries are clearly the responsibility of the project manager, who has some measure of authority over them. Elements that lie outside the project's boundaries are subject only to the manager's influence; by definition they are not the manager's responsibility.

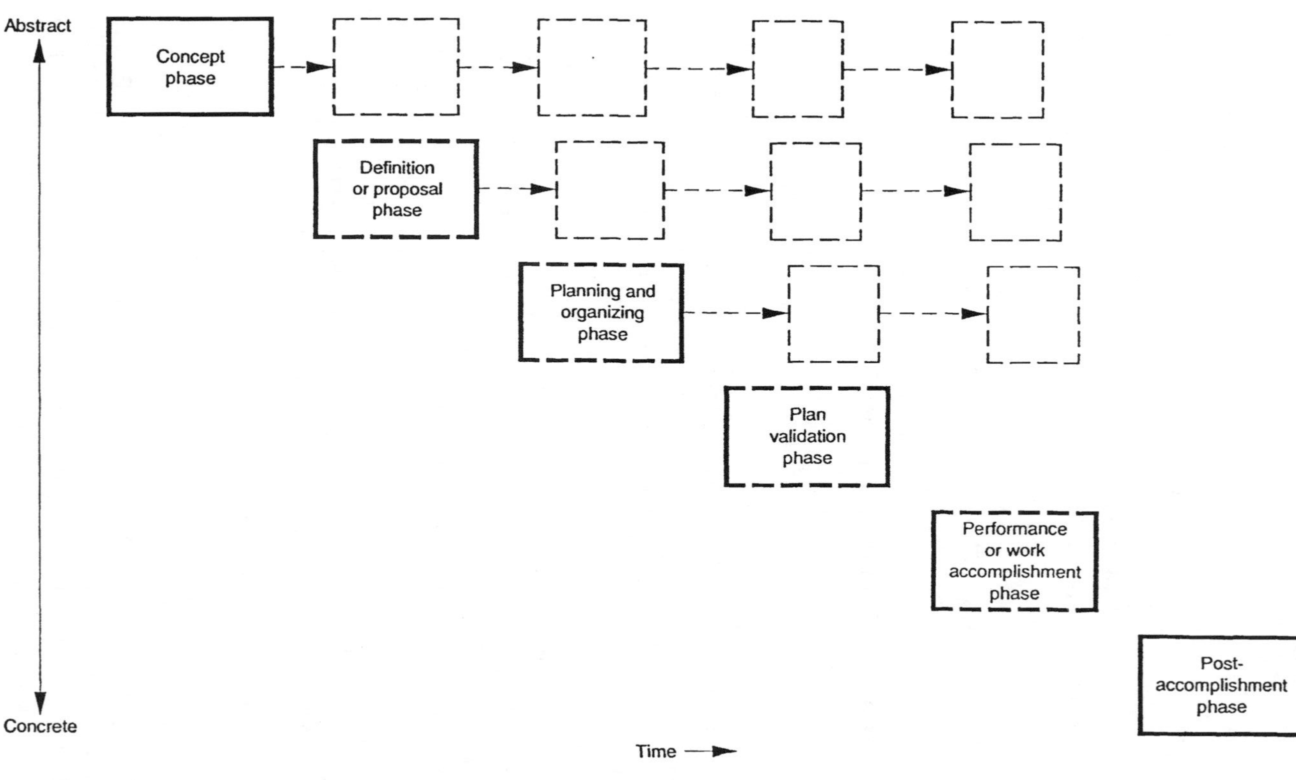

Figure 1.2. Phases of a project, showing possible overlaps and iterations.

While information and resources must generally flow back and forth across project boundaries in order for the project to succeed, project managers may have very limited direct control over these vital exchanges. Accordingly, they must use their insight and diplomatic and persuasive powers to ensure that their projects are not strangled by inappropriate flows. That is, project managers must cause flows to happen without being able to command that they happen. Many times this task requires project managers to negotiate with their counterparts on the other side of these boundaries. The art of negotiation is so important in project management that we devote an entire chapter to it, Chapter 9.

B. Project Inputs

The information, materials, and resources that flow into a project are called inputs and significantly influence how the project will be conducted. These items are discussed in this section.

1. Scope of Work

A scope of work is a statement of project objectives and major constraints, including schedule and budget. Sometimes it is quite detailed; other times it is so poorly stated that the project cannot proceed until it is refined. In any event, the scope of work is a major input to the project.

A scope of work may originate with the customer or it may originate with the organization that will perform the project. The latter may seem strange, but often the would-be customer asks a would-be project organization to propose a project. Although the project organization may have proposed the scope of work to the customer originally, it is only a proposal at this early stage. When it comes back as a contract, it is really the customer's. No project manager should think that the scope of work can be changed without obtaining the customer's prior approval, even if the project organization submitted it as a proposal in the first place.

The project manager should also recognize that work statements described in proposals are often modified by the customer unilaterally or via negotiation with the would-be project organization before they are issued for performance. It behooves project managers to read, nay study, their scopes of work before beginning their projects. And they should keep them handy for future reference (or memorize them). Otherwise, they are prone to working on their projects under false conceptions.

2. The Marketer

Some project organizations are part of larger entities that have marketers who call regularly on their customers. Such people can be invaluable sources of information about political and fiscal climates in customer organizations. Project managers should communicate regularly with their marketers in order to keep abreast of changes in customer feelings long before they surface formally in change orders or suspensions.

3. Contract Terms

Contract terms are also project inputs. They may be either innocuous or consequential, depending upon how well they describe how the project must proceed and the mutual responsibilities of the customer and project organization. Often they are written in a standard format by the customer and applied indiscriminately. Sometimes standard terms are unfair or unsatisfactory to the project organization because of the project's nature. In these cases, they should be revised or removed through negotiation *before* the contract is signed. Also, certain projects need contract terms that are nonstandard and that will appear in the contract only if the project manager sees that they do. Project managers should study the terms of their contracts for the work they will do in order to ensure (1) that contract terms will not so hinder them that they cannot perform their projects and (2) that contract terms will induce the other parties to behave in ways that will aid them in performing their work.

The following contract clauses are usually of concern to project managers when their projects are done under contract for other organizations:

1. *Deliverables and delivery arrangements.* Most contracts specify how and where products are to be delivered, and the project manager needs to plan and act accordingly. Failure to conform to contract delivery requirements can result in default and loss of payment for work that is otherwise correct.
2. *Warranty requirements.* Some contracts specify that the project team or its parent organization will warrant the work done on the project, which means that any defects that show up after the project is complete must be made good by the project organization. It may even have to accept responsibility for any consequential damages that arise because of defects. Prudent project managers take warranty requirements into account when assessing risks associated with alternative project approaches.
3. *Allowable costs.* Contracts often define what costs the customer will pay for. In order to be reimbursed for costs incurred on the project, project managers need to limit project expenditures to allowable types and ensure that expenditures are classified correctly when they are first recorded.
4. *Changes.* Contracts may specify how changes to the scope of work, the schedule, and the budget will be processed, and project operating arrangements must be consistent with any such provisions. Project managers must estimate the amount of "change traffic" to occur and plan to accommodate it properly.

5. *Subcontracts.* Contracts may limit the use or selection of subcontractors, and project managers must plan their work in light of such limitations. In particular, they must allow appropriate time to follow any procedures prescribed concerning subcontractors.

   Also, a project organization is as responsible for the work of its subcontractors as if it did their work itself, even if its customer approves, or even mandates, the selection of the subcontractors. This means that poor, tardy, or overbudget work by a subcontractor may not be used to excuse a project's inadequate performance. Moreover, if a project is having difficulty with the work of a subcontractor, the project is obliged to inform its own customer of the difficulty and what it is doing to correct the situation.
6. *Vendor approvals.* A contract may require a project to obtain prior approval of any vendors that it uses. If so, the project manager must allow sufficient time in the project plan to obtain these approvals.
7. *Security requirements.* Occasionally, contracts specify practices designed to provide security to the work itself or to the customer more generally. In these cases, project plans must allow sufficient time and resources to comply with these specifications, whether they concern security clearances for personnel, protection of proprietary information, access to controlled sites, or compartmentalization of work.
8. *Excusable delays.* Some contracts provide schedule relief for delays caused by excessive bad weather, strikes, riots, acts of God, etc., and some do not. If a project's parent organization has both types of contracts at the same time, those that do not provide such relief get first claim on scarce resources, and managers of these projects need to be sure that they get their due. At the same time, managers of projects that are provided relief from such delays need to ensure that they are not abused and that their customers do not experience un-

due delay as the parent organization tries to minimize adverse consequences overall.

9. *Customer-furnished information and property.* Projects often depend critically upon timely, correct information or property from the customer. Yet many customers do not perform this supporting role properly. Contracts need to address this potential difficulty in two respects. First, the contract needs to identify the information and materials that the customer is to provide. If not all such items can be listed at the outset, then they can be added to the contract by change order or amendment when they are known. Second, the contract needs to provide schedule and budgetary relief to the project if customer-furnished items are not provided as promised. Budgetary relief needs to recognize both the cost of inefficiencies caused by interruptions and the overhead costs that continue when work is interrupted and delivery dates are extended.
10. *Inventions, patents, and copyrightable materials.* A clause concerning inventions, patents, and copyrightable materials identifies the ownership of any of these items developed on the project and specifies any responsibilities that members of the project team may have to the customer regarding documentation, reporting, and assignment of rights. Both the customer and the project manager's own organization look to the project manager to ensure proper documentation and reporting.
11. *Key personnel.* A key personnel clause requires the project to be staffed in a particular way. If a project organization has a key personnel clause in one of its contracts and wishes to substitute other personnel, it must obtain prior approval from the customer. Such approval may not be easy to obtain, and caution is urged in agreeing to such clauses in the first place.

    Key personnel clauses are generally not enforceable against the will of the personnel, since they can be construed to be bonded servitude and thus against public interest. Thus, a key personnel clause can jeopardize a project if a named individual chooses to leave the project and the customer will not accept a substitute.

12. *Overtime and shift premiums.* Occasionally, in order to avoid waiting in a backlog queue, a customer agrees to pay premiums for overtime or shift work. In this event, the project manager must arrange for work that is eligible for premium reimbursement to be done on premium time; otherwise, the premium cannot be collected. Similarly, work that is not eligible for premium reimbursement should be done on straight time; otherwise, premiums will have to be paid out of overheads and profits rather than be expressly reimbursed.
13. *Not-to-exceed figure.* A not-to-exceed (NTE) figure, sometimes called a limit of customer obligations, sets an upper bound on what the customer will pay. An NTE figure is appropriately set above a target cost when the customer intends to bear the risk of an uncertain situation and also wants to limit its financial exposure.

    When an NTE figure is used, the contract should be in terms of "best efforts" toward specific deliverables rather than the deliverables per se. Specific deliverables give the benefits of a good outcome to the customer while placing the burden of adverse outcomes on the project, hardly a symmetric situation. And the situation is exacerbated if the customer can cause difficulties without penalty (see item 9 above).

    If a customer suggests using an NTE figure with specific deliverables, then the project should request an appropriate fixed-price contract instead. These contracts allow the project to keep any benefits that result when things go well to compensate for taking the risk that things may not go well.

    Even when an NTE figure is used appropriately, the project manager must keep track of target and actual costs plus amounts approved via change requests to ensure that the total does not exceed the NTE figure. A higher NTE figure must be negotiated if necessary to cover the target and actual costs and all approved change requests before accepting a change request that would cause the NTE figure to be exceeded. Otherwise, only the lower, insufficient NTE amount will be paid.

14. *Deficiencies and defaults.* When project work does not meet its contractual requirements, the project is either deficient or in default. Deficiencies refer to situations where the customer accepts the work nevertheless, perhaps with an adjustment in price. Defaults refer to situations where the customer rejects the work. When default occurs, the customer makes no further payments, and the project may be unable to recover its expenses. Project managers need to ensure that their projects do not default. This means that they must pay attention to contractual terms as well as to their scopes of work.
15. *Terminations.* A customer may terminate project work either for the customer's convenience or because the project is in default. The previous paragraph describes the consequences of termination by default. A termination-for-convenience clause can provide funds to allow the project to put its records and equipment in order and to pay the salaries of its personnel for a short time, pending their reassignment. Astute project managers make sure that termination provisions are appropriate.

    Toward the end of a project, a termination payment might exceed the cost of finishing the project. In this case, the customer will simply let the work be finished rather than incur the more expensive termination payment. Yet the project staff are likely to know that the work is no longer wanted and may slack off. If they do, default may result. Project managers need to ensure that such an adverse outcome does not occur.
16. *Disputes.* In the absence of contract language to the contrary, disputes between a project and its customer not resolvable amicably will be litigated, generally an expensive process tending to favor the party with the greater ability to withstand the cost. Arbitration is an

alternative that typically reduces both the total cost and the advantage of the party that can better withstand the cost. Many project managers have arbitration specified in their contracts in order to obtain these benefits.

Whether litigation or arbitration is used to resolve disputes, project managers may also benefit from terms that require the losing party to pay the litigation or arbitration costs of the prevailing party. Such terms tend to reduce the likelihood of nuisance suits and promote prompt resolutions.

Contract terms notwithstanding, project managers should negotiate differences whenever possible in order to minimize the need for litigation or arbitration. Timely negotiation of issues saves everyone loss, frustration, and hard feelings (see Chapter 9).

The preceding discussion on contracts implies that project managers are performing work for an external organization, but this need not be the case. Even when projects are done for one's own organization, project managers need to work under conditions that help rather than hinder their jobs. Thus, even here project managers need to negotiate the terms and conditions of their assignments before agreeing to take them.

4. Organization Policies

Organization policies are important inputs in most projects because they guide the way work is performed. For example, they may mandate certain types of reviews and approvals before work proceeds from one stage to the next. They may specify how subcontracts are handled; how support services, e.g., typing and accounting, are provided; how personnel are assigned; and how all-important client contacts or other relationships are maintained. Sometimes they are written down; often they are not. They are nevertheless important influences on project managers' behavior, and they need to know how to succeed in their presence.

5. Project Personnel

Among the most important inputs for project managers are the personnel who are assigned to their projects. They bring technical knowledge and skills, interests, aptitudes, and temperaments. Most projects need a mix of these characteristics, as described in Chapter 6. The project manager should try first to identify the kinds of talents needed and then to secure them, not just tell the personnel department that the project needs two X-ologists, one Y-ologist, and four Z-ologists.

6. Material Resources

Material resources comprise the facilities and equipment that can or must be applied to the project. Often they limit how the project can be achieved. Or, if they have unusual capabilities or versatility, they may enable particularly effective or efficient approaches which are not always used. In either case, project managers should understand how the material resources available to them are likely to affect or facilitate their projects. They should be identified as to

a. What the customer shall furnish
b. What can be obtained from within the project manager's own organization
c. What shall have to be obtained elsewhere, including likely sources

7. Information

Available information is a project input and influences how a project is planned and executed. Information may be technical, economic, political, sociological, environmental, and so forth. It can come from other members of the organization, from customers, from subcontractors and vendors, from third parties such as governmental agencies, and of course from both open and restricted literature.

The quality and quantity of information which the project organization obtains, either on its own initiative or on the initiative of others, will have major impact on the amount and type of work that can be and must be done. Project managers and their teams should cultivate their information sources and use them advantageously. Little else will be as cost-effective if the information exists.

8. Upper Management

The project manager's own upper management is a source of input for many projects. Upper management may view a project as a vehicle for accomplishing its own agenda, even if the project's customer is outside the project's parent organization. In effect, management is another facet of the total customer; the project manager should act accordingly.

Occasionally, project managers find that their upper managements and their bona fide customers have conflicting objectives or constraints. These conflicts are untenable for project managers, and they must then negotiate changes that will result in compatible objectives and constraints.

C. Project Outputs

The information, materials, and resources that flow from a project are called outputs, discussed in this section.

A project can be visualized as a processing machine. The inputs from Section II.B are processed by the project activities to produce the project outputs:

1. Deliverables
2. Internal information
3. Experienced personnel
4. Working relationships

1. Deliverables

Deliverables are the visible products of the project, the items that the project is supposed to produce. They may include tangible items such as hardware or a structure, or information such as instructions, an analysis, a report, drawings and specifications, a design, contract documents, or a plan.

2. Internal Information

Internal information consists of the increased knowledge base that the project staff generate as a result of doing the project. This information can be invaluable to the organization in a future project, whether for the existing customer or another customer. Above all else, it includes the actual durations and resource costs of the individual pieces of the project, which can be used in estimating durations and costs on future projects. Internal information may also be about the customer, about subcontractors, about production conditions, and about the environment as well as about the project itself. The information may be in the form of memoranda, report drafts, project instructions, project standards, check prints, and so forth.

Since the project is not required to deliver such internal information, it is often neglected. Astute project managers, however, will retain such information whenever they can in order to build their knowledge base for future assignments.

3. Experienced Personnel

A major output of every project is experienced personnel. Whether the experience is in fact beneficial or harmful depends of course on how well the project is conducted. If it is conducted satisfactorily or even outstandingly, project staff will grow and develop and generally enjoy their experience. They will be both more skillful and willing to work with the project manager in the future, which is a plus for the project manager. On the other hand, if the project is conducted poorly, project staff will probably not enjoy their experience even if they do grow and develop. The odds in such circumstances are that the staff will prefer not to work with the project manager if they can avoid it, which limits the manager's options, if not his or her effectiveness, in the future.

Table 1.1. Potential Project Interfaces

Within the organization:

- Accounting office
  - accounts payable
  - accounts receivable
  - project accounting records
- Administrative or business manager
- Computer facility
  - application programmers
  - supervisors
- Contract administration
- Drafting department
  - drafters
  - supervisors
- Engineering disciplines
  - engineers
  - supervisors
- Equipment facility
- Estimating department
- Field offices
  - supervisors
  - technicians and other staff
- Insurance office
- Laboratories
  - supervisors
  - technicians
- Legal department
- Library
- Logistics specialists
- Main office
  - central engineering
  - inspector general for projects
- Marketing department
- Material department
- Other offices
  - office head
  - project managers
- Other project managers
- Personnel office
- Printing department
- Public relations department
- Purchasing department
- Quality department
- Technical project office
  - supervisor
  - technical project officer
- Technical specialists
- Transportation office

With the customer's organization:

- Accommodations office
- Accounts payable
- Contract administration
- Drafting department
- Equipment supply office
- Field staff
  - supervisors
  - technicians
- Industrial relations department
- Legal department
- Public relations department
- Purchasing department
- Quality department
- Receptionist
- Safety department
- Scientific disciplines
  - discipline managers
  - principal investigators
- Secretaries

Table 1.1. Continued

Security department	Transportation office
Subcontract administration	Travel office
With subcontractor and vendor organizations:	
Accounts receivable	Project manager
Contracts administration	Public relations department
Crew chiefs	Purchasing department
Equipment supply office	Quality department
Insurance office	Sales department
Legal department	Storekeepers
Materials department	Technical specialists
Operators and technicians	Transportation department
Pricing department	Workshop
With third parties:	
Auditors	Libraries
Banks	Media
Competitors	Planning and zoning agencies
Data bases	Politicans
Equipment rental firms	Pressure groups
Express and freight-forwarding companies	Regulatory agencies
Family members	Surveyors
Governmental agencies, including permit issuers	Utilities
Land owners	Tax assessors
Learned societies	Travel agents

4. Working Relationships

Another project output is the set of working relationships that the project manager develops with other internal departments and with external organizations. Like the quality of the project staff's experience, these relationships may be either positive or negative and they will bode either well or ill for the future. Project managers should take care to develop the kinds of relationships on current projects that will serve them well on later projects.

D. Project Interfaces

A typical project has literally dozens of interfaces across which information and deliverables flow. These interfaces are with various departments and offices in the project's own organization, various functional elements in the client's organization, different elements in subcontractor and vendor organizations, and third parties, such as regulatory agencies, auditors, and so forth.

A major duty of every project manager is to ensure that all project interfaces function so that necessary information and materials are properly transmitted. The first step in managing these many interfaces is simply to list them and to assign each one to a project staff member who will keep it functioning. To help in this regard, Table 1.1 lists representative interfaces for reference.

The next step is for all project staff members to establish face-to-face or telephone contacts with their interface counterparts and to make their existence known. They should exchange (or at least begin to exchange) information with each of them on their respective needs, wants, and expectations. Each staff member should report back to the project manager the agreements that they and their counterparts have made about the nature, form, and frequency of their future exchanges. The work implied by these future exchanges should then be factored into their project plans, which are discussed in Chapter 3.

As information and deliverables begin to flow across the interfaces, the interface-keepers should verify that they are serving their intended purposes. This responsibility may require some deliberate follow-up activity, for recipients may not report whether the items satisfy their needs and expectations and whether additional work or information would be helpful.

2

Roles and Responsibilities of the Project Manager

Project management involves a set of duties that must be performed and are no one else's prime responsibility. We have organized these duties into six roles and responsibilities. Depending upon the project organization, the first five may be done by the project manager directly or they may be done by others under the project manager's supervision. The sixth one, inherent duties, can be done only by the de facto project manager.

I. ENSURE CUSTOMER SATISFACTION

The single most important responsibility of the project manager is to ensure customer satisfaction. If the project is successful in every respect in terms of meeting its stated objectives, schedule, and budget but the customer is somehow not satisfied, then the job was not done well enough. Such dissatisfaction could arise because the customer had a preconceived, but unspoken notion of what the outcome should be, because the customer obtained new information that caused a revision in priorities, or because the customer had a change of mind.

The project manager can learn about and keep abreast of the customer's needs and expectations by practicing some simple procedures:

1. Confirm key issues during the course of the project so that the work done is adjusted to meet current needs and expectations.
2. Develop a friendly intelligence system within the customer's organization to obtain early warning signals of changes in emphasis, priority, etc., and prepare to respond appropriately to them.
3. Reread the contract several times during the project and attend to all its requirements, including administrative aspects, so that no needs or wants are left unfulfilled.
4. Keep the customer informed and up-to-date so that he or she is prepared for the eventual results and has an opportunity to influence the way that they are developed and presented.

If there is a written contract, it may seem strange that it is necessary to be in touch continually with the customer. However, a project situation is seldom straightforward, and the contract may not reveal all that needs to be taken into account. For example,

1. The customer may not have stated the requirements clearly or completely. Contact is necessary to determine the true objectives and constraints.
2. The contract or statement of work may not reflect the needs of all the elements of the customer. Contact is necessary first to identify all the interested elements and then to identify their needs.
3. The customer may have constraints or a hidden agenda which cannot be put in the contract or scope of work but whose existence controls the range of useful project re-

sults. Contact is necessary to ferret out these unwritten limitations.

The entire project team can develop information about their customer as they do their day-to-day work on the project, and the project manager should guide them in fulfilling this role. Many inexperienced project personnel need to be instructed regarding these duties.

Finally, if it should become clear that the customer cannot possibly be satisfied, then the project manager should

1. Be sure that the contract is satisfied to the letter
2. Be sure that the customer pays all invoices due
3. Terminate the relationship

II. DIRECT AND CONTROL ALL DAY-TO-DAY ACTIVITIES NECESSARY TO ACCOMPLISH THE PROJECT

A project does not run itself. Someone must direct and control it, and that someone is the project manager. The project manager is responsible for directing and controlling all the day-to-day activities that are necessary to accomplish the project.

This responsibility does not mean that project managers cannot delegate the direction or control of specific activities to others. On the contrary, they may indeed delegate, but the *responsibility* for the direction and control is still theirs. If the delegates do not perform adequately, the project managers are responsible for the inadequate outcomes and for repairing any damage done.

Good project managers arrange matters so that their teams avoid troubles and have appropriate back-up plans and resources for risky situations. By their foresight, their projects run more or less smoothly. The more successful they are in this regard, the less likely anyone will be aware of their effectiveness. As a result, they may not be fully appreciated. Only sophisticated observers will recognize the high quality of their efforts.

A project manager may find from time to time that a boss, a salesperson, a functional manager who supplies some staff, or even a customer is trying to direct part of the project. When this happens, the project manager must firmly rebuff the interference. A project manager may well listen to advice from these people but should not let them direct project staff members. Otherwise, there is a grave risk that the project will be seriously damaged overall. This occurs because the would-be intervenor invariably lacks understanding of the entire project and its interdependencies. While tinkering with one part, the intervenor upsets the project's overall balance even if well-intentioned.

Directing and controlling day-to-day activities is both time-consuming and sporadic. Its sporadic nature makes it difficult to schedule completely. If project managers are not careful, they may leave too little opportunity in their own work plans to accommodate the many conversations and informal meetings that are necessary to direct and control their projects. As a rule of thumb, they should commit to formal meetings and appointments at most half of their time. The other half will be filled with directing and controlling day-to-day activities.

III. TAKE INITIATIVES AS REQUIRED IN ORDER TO ACCOMPLISH THE PROJECT

The project manager is the chief initiator on the project. From time to time he or she will face and have to resolve problems that no one could foresee. There will always be revolting developments and they must be expected. Action is required when they happen.

In all likelihood, project managers will be unable to resolve every revolting development alone. They may need to consult an expert but not know to whom to turn. When this happens, they can call any person whom they think is likely to know more about the problem than they do. The chances are that this first contact will not be able to provide all the help needed but can probably give one or more leads to people who have more expertise. They can continue this chain a few more times. The chances are very good that by the fifth phone call they will be connected to someone who will be a suitable expert for the problem. (This process is known as the "five-phone-call phenomenon.")

If the bottom drops out and the project manager gets a sinking feeling, he or she should not just stand around and act like a bump on a log. That would only cause others to do so, too. The project manager must take the initiative and attempt to solve the problem as quickly and as orderly as possible.

IV. NEGOTIATE COMMITMENTS WITH THE CUSTOMER

The project manager is responsible for negotiating the commitments that the project team will perform for their customer. As explained in Chapter 9, this does not necessarily mean that the project manager is an active negotiator for all the commitments, but he or she is responsible for the team's negotiations. In the end, the project manager is responsible for fulfilling whatever is negotiated, and therefore has a vital interest in the outcome of the negotiations and must be in agreement with it. If not, then the project manager is not prepared to manage the project.

Sometimes a project manager is assigned to a project after some commitments have already been made to the customer. The project manager's first duty in these circumstances is to evaluate the situation and determine if the project team can meet the commitments. If they can, well and good. If they cannot, then the project manager must tell management and/or the customer what problems exist and negotiate relief from unreasonable objectives and constraints.

If the commitments cannot be changed, then the project manager should document the envisioned difficulties to management so that they will know what to expect.

In most projects, negotiations continue during the work. Most customers recognize that nothing is certain and that changes may be necessary once the facts of the situation emerge. Depending upon the degree of specificity of the contract or arrangement, formal changes in objectives, schedule, budget, staffing, strategy, and methods may have to be renegotiated.

Negotiating is discussed further in Chapter 9.

V. ENSURE COLLECTION OF THE FEE

The project manager is responsible for ensuring that any payments due are collected from the customer. In some organizations, this task is efficiently and effectively handled by someone other than the project manager. However, the project manager should always know how well the customer is meeting contractual commitments of payment. If payment should lag seriously, the project manager may stop work on the project, both to avoid incurring further expenses without reimbursement and to induce the customer to pay for work already done.

Also, the project manager must ensure that the project team fulfills all its contractual commitments properly so that the customer will have no reason to withhold payment. This means that the project manger and key team members need to know not only the scope of work but also administrative details and the contract's terms and conditions.

VI. INHERENT DUTIES

This section lists nine duties that we believe are inherent duties of the project manager. Putting it another way, the de facto project manager is the one person who performs *all* these duties on a project.

If these duties are divided among two or more people, then the project manager's role and responsibility are also divided. This means that the project staff and all others who deal with the project will be confused about who the project manager really is. Also, the "alternative" project manager might be consulted and notified when the nominal manager should be. The staff can thus receive instructions contrary to those that the project manager would give and he or she will lack important information needed to manage the project.

A. Interpret the Statement of Work to Supporting Elements

Interpreting the statement of work is the first opportunity for the project manager to show leadership. It is a critical step. When done well and timely, it is a good motivating tool for the entire project team. It is the key to everyone on the team having a common understanding of the objectives, constraints, and major interactions on the project.

B. Prepare and Be Responsible for an Implementation Plan

The project implementation plan must be the project manager's own plan. While the project manager should involve others in preparing the plan, both to obtain their expertise and to allow them to influence the plan so that they can be committed to it, there should be no mistake that it is the *project manager's plan*. Everyone should consider a violation of the plan to be a violation of the project manager.

To convey this sense of ownership, the project manager must make the difficult planning decisions personally and behave consistently with them. Otherwise, project team members may believe that the project manager does not understand the difficult matters or does not support them.

If such an attitude develops, team members may feel free to deviate from the plan as they choose, and the entire project may suffer.

If someone becomes the manager of a project in mid-course, then the new manager should develop his or her own plan and publish it promptly. Even if the new manager accepts the plan left by the previous manager, the new manager must affirm that it is now his or her plan and commit to it.

C. Define, Negotiate, and Secure Resource Commitments

The project manager should define, negotiate, and secure resource commitments for the personnel, equipment, facilities, and services needed for the project. By negotiating for resources, the project manager establishes his or her role in the eyes of all those who provide resources and support to the project.

Resource commitments should be as specific as possible in order to verify that they are appropriate and that they are being kept, and they should be obtained from those organizationally able to make commitments. Thus, resource negotiations will be primarily with those who can supply the resources, but the person who resolves priority conflicts among projects (hereafter called the director of projects) will also be involved if available resources cannot serve all project needs in a timely way. If a project manager must yield resources to a competing project, then he or she should seek appropriate schedule, cost, or scope relief from the director of projects to compensate for yielding.

Discussions of resource commitments should address the possibility that a person supplying a resource might be asked to reassign it in order to solve an emergency elsewhere. The project manager and each resource supplier should explicitly agree that such a reassignment will not be made without the supplier first notifying the project manager of the request and allowing the latter to appeal the proposed reassignment to the director of projects. If the resource is to be reassigned despite the appeal, then the project manager and the director of projects should negotiate the timing of the reassignment and any specific schedule, budget, or scope changes needed to accommodate the loss of resource.

D. Manage and Coordinate Interfaces Created by Subdividing the Project

As explained in Chapter 3, projects are typically broken into subprojects that can be assigned to individuals or groups for accomplishment. Whenever the main project is broken down this way, the project manager must manage and coordinate the high-level interfaces that are formed by subdividing the work. (Intermediate and lower level interfaces are managed and coordinated by the task leaders who create such subdivisions.)

Interface management and coordination consist of making sure that each leader of a subproject provides whatever another subproject leader needs and does so in a timely way. This may require intervening in a subproject leader's plan, since meeting interface requirements may not be the leader's own highest priority task. Likewise, the project manager may need to persuade another subproject leader to moderate his or her requirements because meeting them may place undue strain on the subproject providing the needed items or information.

Sometimes subproject interfaces can effectively be managed and coordinated by holding a series of separate meetings with the individual subproject leaders. At other times, however, best results are obtained by having all the subproject leaders discuss their needs and constraints together at a project meeting. A project meeting enables all the participants to contribute toward resolving their common difficulties without their views being filtered through the project manager. Moreover, it generally facilitates imaginative solutions and a full discussion of side effects, which are less likely to occur in a series of individual discussions.

By establishing interfaces where they can be readily described in advance, with communications as straightforward as possible, project managers can minimize the time they spend in interface management and coordination, making time available to do other tasks. Having said this, however, project managers should not shrink from resolving disputes that arise from the way they subdivide their projects. Resolving such disputes is a major factor in establishing who is in charge, and performing this responsibility provides project managers considerable authority over their projects.

Other guidelines for subdividing or partitioning work are given in the discussion of work breakdown structures in Chapter 3.

E. Monitor and Report Progress and Problems

The project manager is responsible for reporting progress and problems on the project to the customer, the boss, and all others who need to know. This responsibility should not be delegated lest the outsiders start to go directly to team members with their concerns.

Team members should be instructed to refer inquiries and requests from outsiders to the project manager if they involve issues not directly within their own personal control. Also, when team members do respond to outside inquiries or requests, they should promptly notify their project manager of their actions.

In order to report progress and problems in a timely way, project managers must monitor their projects and not merely wait until others bring them news. They must be aware of what is going well and what is going awry and should seek these kinds of information from their project staff. They may have to persist in uncovering unhappy news, for many people are reluctant to give it. If, however, they explain that they prefer early warnings to late surprises, help their staffs surmount their difficulties, and do not shoot their messengers, then they stand a good chance of getting the information that they need first to keep their projects on course and second to keep outsiders properly apprised of their situations.

Details of the process of monitoring are discussed in Chapter 4 on control techniques.

F. Alert Management to Difficulties Beyond One's Control

Occasionally a project manager finds that the only way to relieve a bad situation is to get help from outside the project. When this happens, the project manager has a difficulty beyond one's control. Suffering the difficulty in silence will not be rewarded. The project manager's management or customer must be alerted so that other resources can be brought to bear, constraints can be relaxed, or project objectives can be adjusted.

It is not sufficient just to mention the difficulty in passing. The discussion must be an overt act and should be confirmed in writing. If the management or customer says that they will take specific action to resolve the difficulty, that too should be in writing, together with the time when they say it will be done.

G. Maintain Standards and Conform to Established Policies and Practices

The project manager must set and maintain the standards that will govern the project staff members. Where pertinent policies and practices have already been established, the project manager is responsible for seeing that the team conforms to them.

Whether or not project managers take overt action, most project staff will take their cues from them regarding acceptable standards of behavior and performance. The occasional maverick who resists following these norms should be faced directly on issues that are important in order to build a group or team approach to the project. The performance of the regulars can be undermined by allowing a maverick's performance or behavior to go unchecked on an important issue.

At the same time, project managers should not impose needless constraints or standards. Nor should they act in ways that appear arbitrary or capricious. If they are to win and maintain the respect of their team members, their actions should be seen as helpful and reasonable, not handicapping or highhanded.

H. Organize and Present Reports and Reviews

The project manager is responsible for organizing and presenting reports and reviews to the customer and to management. He or she is the focal point for the team and should be seen as such.

This duty does not imply that project managers must either prepare or present 100% of their project reports or reviews. But they are responsible for them and should orchestrate them, send written reports under their cover letters, and be the first and last speaker and the glue that holds a review together.

In preparing for a review, the project manager should ensure that the presenters understand the limits of their authority to answer questions from members of the audience. They may answer questions about what they said or meant, but they should not answer questions about the significance of their work to elements of the project for which they are not responsible or to the project at large. Rather, presenters should refer such questions to the project manager (even if they know the answers). This behavior will strengthen the project manager's role as project integrator and will help cement the relationship between the audience and the project manager.

The project manager who receives a question in this fashion must decide how to respond. The manager may choose to answer the question personally; to refer it to an expert staff member, even back to the presenter; or to promise to get an answer back to the questioner soon. The latter is useful if the requisite expert is absent or if a study or caucus is needed before answering. In any case, the project manager should not feel inadequate if unable to answer all questions immediately. The project manager is not expected to know every last detail but is expected to have the resources to handle pertinent issues. (Impertinent questions should be politely rebuffed on the spot.)

I. Develop Personnel as Needed to Accomplish the Project

It is a rare project indeed where all the talents needed to do the work are present in the staff available. The project manager must thus train them and compensate for their shortcomings. A project manager may, for example, have to teach staff members how to plan their work, interface with others, report their results, and generally how to function as a project team member. The manager may also have to teach them how to practice their specialties under adverse circumstances by showing them how to provide for risk.

Developing project personnel can produce two subsidiary benefits. First, project personnel who know that their project manager is interested in their development are more likely to acknowledge their shortcomings early and seek to repair them before they cause difficulty. Otherwise, project personnel may be inclined to hide shortcomings, hoping that they will not be discovered, and thereby cause problems that are harder to avoid or fix. Second, project personnel who know that their project manager is interested in their development are more likely to feel a measure of gratitude and loyalty toward the project manager. These feelings can heighten commitment levels and may thus help the project succeed in instances where it otherwise would not.

Whereas project managers can delegate parts of the tasks described in Sections I through V, they dare not delegate their inherent duties. To do so is to confer upon others part or all of the role of project manager and undercut their own ability to manage. In this event, they will be lucky if their projects turn out successfully.

If one is assigned as a project manager and cannot perform all nine inherent duties, he or she should rebel! If one is asked to take over a project in trouble and cannot perform all nine inherent duties, he or she should rebel! Perhaps this explains why the project floundered in the first place.

3

Planning Techniques

It was mentioned in Chapter 2 that an inherent duty of a project manager is to prepare and be responsible for an implementation plan. This duty lies at the heart of project management and warrants its own chapter. The duty includes defining tasks, making estimates, and preparing schedules and budgets. It also involves assigning overall segments of the plan to individuals, while minimizing interface problems.

I. WHY PLAN?

Before embarking on the discussion of planning techniques, it is appropriate to address the question, "why plan?". Isn't the plan outlined in the project definition or proposal adequate? In general, the answer is that the project definition or proposal plan is incomplete and too superficial to serve as a project *management* plan. It is typically prepared to sell the project and does not address many elements needed to manage the project. And it may be a success-oriented plan, which is hardly a protection against mishaps or a means of contending with them. Thus, the project manager must develop a project plan in sufficient detail, which is discussed later.

There are still additional reasons to plan:

1. The plan is a simulation of prospective project work, which allows flaws to be identified in time to be corrected.
2. The plan is a vehicle for discussing each person's role and responsibilities, thereby helping direct and control the work of the project.
3. The plan shows how the parts fit together, which is essential for coordinating related activities.
4. The plan is a point of reference for any changes of scope, thereby helping project managers deal with their customers.
5. The plan helps everyone know when the objectives have been reached and therefore when to stop.

II. DEFINING TASKS AND WORK BREAKDOWN STRUCTURES

Defining project tasks is typically complex and accomplished by a series of decompositions followed by a series of aggregations. The decompositions involve *conceptually* breaking the final project products into smaller simpler pieces which are themselves decomposed into still smaller pieces, and so on until individual realizable pieces are identified. The aggregations involve *physically* combining these smallest pieces to form larger pieces, then combining the latter to form still larger pieces, and so forth until the final project products are formed.

The first step in the decomposition is to identify the overall project requirements and subsidiary requirements, for example, reporting and billing requirements. In some cases, these can be determined solely by rereading the scope of work and the contract and by referring to applicable regulations and to organizational policies and procedures. In other cases, interviews must also be conducted with customers, regulators, and so forth to identify all the project's requirements. A process for determining requirements is suggested in Appendix A.

The requirements serve as input for conceptualizing, defining, and characterizing a compatible set of subproducts that can be integrated to form final products. Subproducts may be physical or informational and may include tooling, staff, and so on, as well as the items that will become the output. Each subproduct has its own required features that enable it to fit or interact properly with the other subproducts in order to satisfy overall requirements. A process for choosing subproducts is described in Appendix A.

Each subproduct can be similarly developed from sub-subproducts, and the subdivision process can be repeated successively many times. Theoretically, the number of subdivisions has no limit. Practical limits exist, however, based primarily on what can be economically obtained in the marketplace.

Once the items at the lowest level have been identified and characterized, they are acquired or built and then validated to be sure that they are correct. They are then assembled or integrated to form the next higher level of subproducts. This process is continued upward until the final products are validated. Integration and validation are also described in Appendix A.

Defining the final and intermediate products and their relationships as just described is called developing a product-based work breakdown structure (WBS). The WBS can be represented in a tree-like chart, similar to an organization chart, that shows the subproducts of each product. Figures 3.1 and 3.2 show portions of a WBS for a spacecraft attitude and articulation control subsystem (AACS).

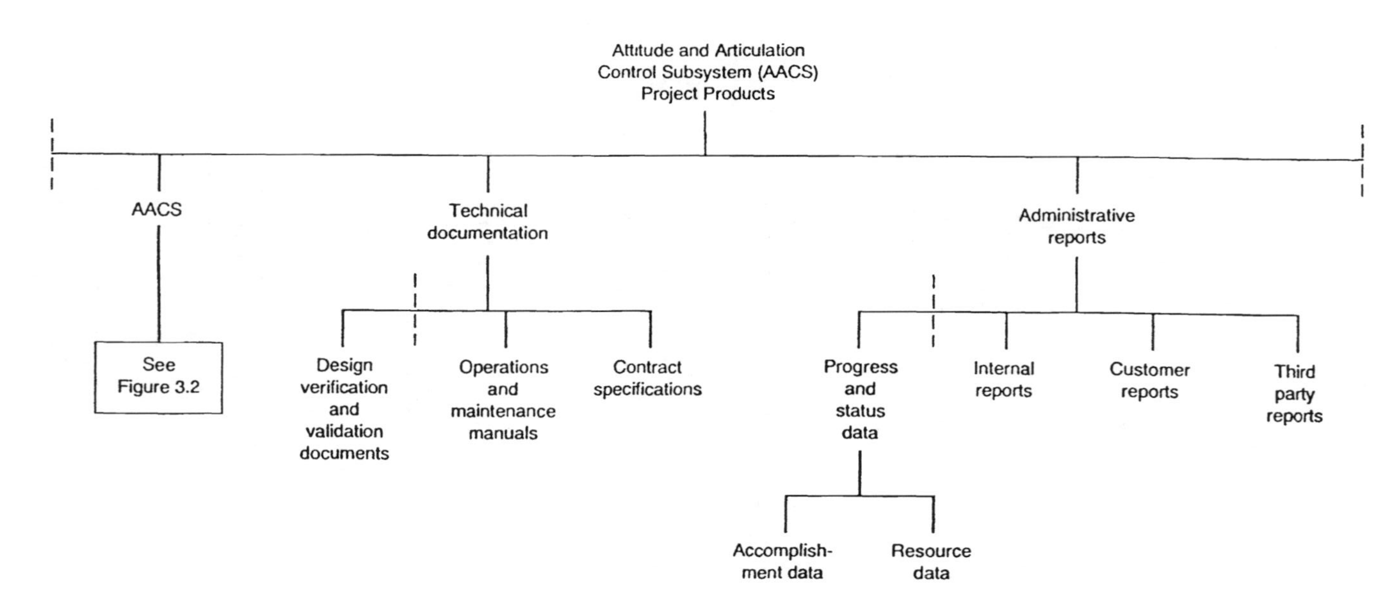

Figure 3.1. The beginning of a work breakdown structure for a spacecraft attitude and articulation control subsystem's project products.

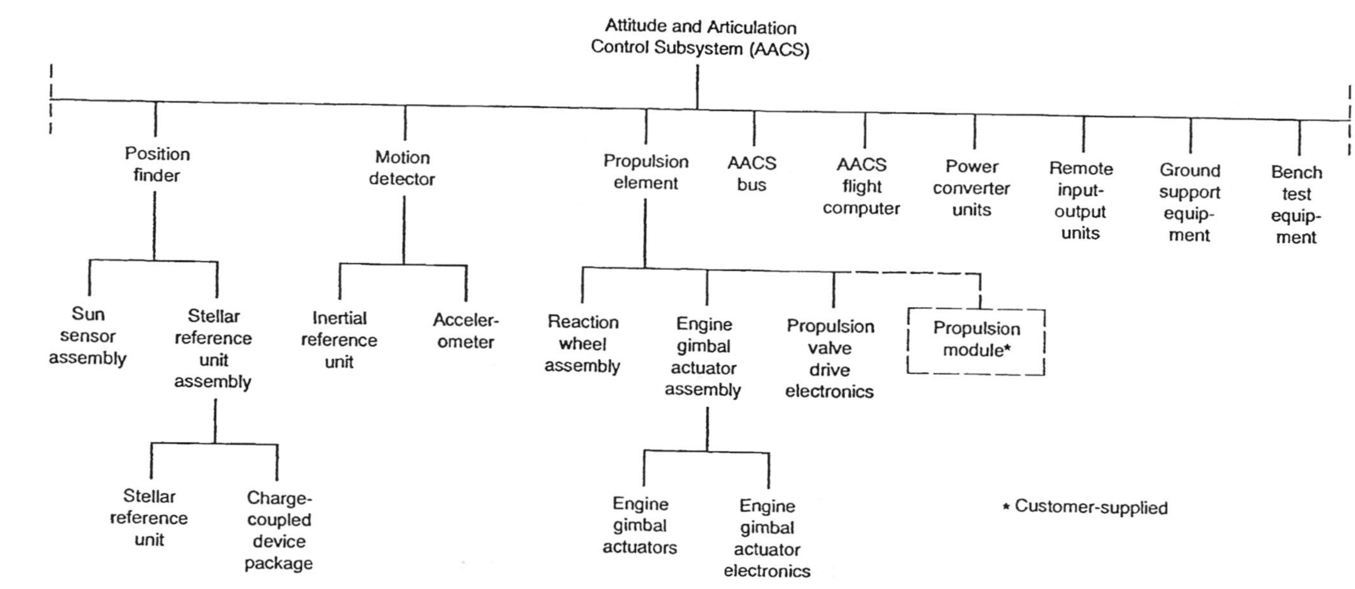

Figure 3.2. A portion of a work breakdown structure for a spacecraft attitude and articulation control subsystem.

Each element in the WBS needs a schedule, a budget, and a responsible individual for its preparation. These individuals should regard preparation of their respective subproducts as (sub)projects and perform all the project manager responsibilities discussed in Chapter 2.

In particular, subproject managers should *negotiate* their products, schedule, and budget *as a set* with their immediate customers. They should also negotiate their interfaces with their sibling subproject managers. Their agreements are tentative, however, until all related negotiations are concluded and found to be mutually compatible.

In defining subproducts and subprojects, awkward arrangements should be recognized and avoided. Usually, confusion is minimized and effectiveness and efficiency are enhanced by observing the following guidelines:

1. *Interfaces.* Each WBS element should have clean, manageable interfaces with its parallel elements. In particular, interface definitions between parallel portions should be easy to describe in advance or to negotiate. An interface should not be put in the midst of a complex or chaotic piece of work.
2. *Verifiability and validatability.* A verifiable element is one that can be shown to have been put together correctly (or incorrectly). A validatable element is one that can be shown to meet its requirements (or not meet them). The WBS should define elements that can be verified and validated as they are formed from their subelements, in order to proceed confidently from one level of integration to the next.

   Occasionally, validation can be achieved only after many levels of integration have occurred. In these cases, one must rely on verification to develop confidence that validation will be realized in the end.
3. *Robustness.* Robustness is the ability to accommodate small to moderate changes in requirements with only small to moderate adjustments of ongoing work. Parti-

tioning is considered robust if future changes affect a small number rather than a large number of WBS elements. In order to devise a robust WBS, the project team must anticipate future changes as well as it can despite the difficulty of doing so.

4. *Organizational fit.* WBS elements that map conveniently onto existing organizations are preferred over elements that span two or more organizations, other things being equal. When a portion of work can be performed entirely by an existing organization, the overall project can reap any efficiencies that the organization has.

   Efficiency (doing things right) should never be obtained, however, at the expense of effectiveness (doing the right thing). If mapping WBS elements to existing organizations threatens effectiveness because it violates other guidelines in this list, then organizational mapping should be sacrificed in favor of more appropriate ad hoc arrangements.

   Examples of ad hoc arrangements are task teams and contractor-subcontractor relationships (within the overall organization). Of the two, contractor-subcontractor relationships disrupt existing arrangements less and are more likely to focus responsibility clearly. However, it requires the willingness of one organizational element to be subcontractor to another, which does not always exist.

5. *Element size.* A WBS element needs to be of such granularity and detail that the person responsible for it can answer the following questions with authority and without ambiguity:

   a. Is the product complete?
   b. If the answer to a is yes, is the work correct?

   This approach means first of all that WBS elements are assigned to individuals who are competent to judge the completeness and correctness of the products assigned to them. And it also means that products are assigned in such a way that these individuals need to check only with those who are responsible for hierarchically related subparts in order to answer the questions.

6. *Subproduct self-sufficiency.* It is often tempting to create upper-level WBS subproducts that are as self-sufficient as possible. In many cases, however, this self-sufficiency forgoes economies that could be realized if certain sub-subproducts or sub-subfunctions were provided once for use in two or more subproducts.

   Accordingly, subproducts and their sub-subproducts should be anticipated as each level of a WBS is developed so that appropriate upper-level utility subproduct items can be defined and provided to lower-level elements. Examples of utility products are data and information storage, computing, communication, structural modules, power supply, and monitor and control.

7. *Standardization.* Once effectiveness is assured and attention is directed toward efficiency, certain standardizations across the WBS can be considered. Examples are standard hardware items, standard report formats, standard software languages, standard test methods, and so forth. Their use may require subproduct–subproduct interfaces to be revised, and it is appropriate to do so as long as satisfactory results are obtained.

   Where institutional standards exist, they may be invoked, perhaps with modification. Where no relevant standards exist, they may be created to serve the project at hand.

8. *Accounting considerations.* Accounting requirements should not control the definition of WBS elements. Modern computers allow cost and other data to be collected and summed in any fashion desired, and no valid reason exists to force a particular WBS structure just to satisfy accounting needs.

9. *All-at-once partitioning.* Whenever possible, the first level breakdown of a given WBS element should be de-

fined all at once, thereby allowing the total set of project subproducts to be defined optimally overall. To proceed piecemeal is to run the risk of a poor WBS.

All-at-once partitioning does not mean, however, that an entire WBS should or can be developed completely at the beginning of the project. Rather, a WBS, like all facets of project plans, should be elaborated and revised as new information becomes available.

These guidelines will not guarantee optimum partitioning, for imagination and skill are needed to conceive the best way. They will, however, help project managers minimize unnecessary difficulties.

Associated with each element in a WBS, regardless of its level, are its inputs and outputs. The outputs at one level are the major inputs to the next higher level. When an element is subdivided into smaller pieces, the person who was responsible for the element before it was subdivided is responsible for coordinating the resulting interfaces, arbitrating interface conflicts among the lower-level pieces, and assembling the pieces to form the larger element.

Inputs also come from parallel or sibling branches of the WBS. Two types of sibling relationships are possible. In one type, no inherent chronologic relationship exists. That is, no sibling depends upon another sibling being developed first. Concurrent development of siblings in these situations enables information to be traded back and forth, which can be advantageous. This type of sibling relationship is shown in Figure 3.2, where AACS ground support equipment and bench test equipment are developed in parallel, quite possibly to their mutual benefit.

Another example of trading information back and forth in Figure 3.2 concerns the AACS flight computer. The computer is a utility whose customers include the position finder, motion detector, and propulsion element. In order to develop both a suitable computer and an optimum AACS overall, potential computer capabilities and the requirements of these diverse customers must be considered together. While the AACS design specifies a flight computer to serve multiple customers, it cannot specify all customer computer requirements in detail; these must await the preliminary design of the customers themselves. Accordingly, the position finder, motion detector, propulsion element, and the flight computer are parallel or sibling products with no one of them necessarily leading the others.

In the other type of sibling relationship, phasing is inherent. For example, progress and status data must be prepared before internal reports, customer reports, and third party reports can be prepared. Phased products are shown as though they were parallel except that they are separated by short vertical dashed lines. Thus, in Figure 3.2, a short vertical dashed line separates progress and status data from internal reports, customer reports, and third party reports.

One who needs an input must negotiate for its delivery in a timely way. And if such a negotiation stalls, then the individual who is just high enough in the WBS to span the affected elements must resolve the difficulty. Thus, in Figure 3.2, if the computer's developer and its customers are unable to agree about priorities or timely transfer of physical or informational products, their disagreement should be resolved by the lowest person in the WBS who spans the affected elements; in this case it is the AACS manager.

To summarize, developing a WBS means negotiating inputs and outputs and responsibilities. Usually, some number of iterations are required to obtain a fully consistent set of arrangements from level to level and across all the interfaces. In many cases, some of the partitioning must await information that will become available later. Thus, the process can be both time-consuming and extended. It should not, however, be shortchanged or treated lightly. Project success depends greatly upon reaching explicit, compatible, and realizable agreements.

III. PRECEDENCE CHARTS

Work breakdown structures show graphically how higher-level products are hierarchically composed of lower-level products that must come first and that certain elements cannot be settled until other elements are settled, but they do not show timing relationships consistently. That is, time runs from bottom to top within segments of the WBS that show hierarchical composition, and time runs from left to right across segments of the WBS that must be completed in a specified sequence. In order to show both of these relationships with a common time dimension, another representation is needed. It is called a *precedence chart.*

A precedence chart is a diagram that shows the sequence and relative timing of all activities on a project at whatever level of detail is desired. As described in Appendix B, it can be created largely by transforming the project's WBS.

One form of precedence chart is a series of boxes or circles, one for each activity, joined by arrows to show the sequence in which the activities must be performed. An alternative, and the form used here, is a series of arrows, one arrow for each activity, joined head-to-tail to show the sequence in which the activities must be performed. Figures 3.3a–f are a precedence chart for the spacecraft AACS development whose WBS is represented in Figures 3.1 and 3.2.

The sequence of activities is dictated by the output–input relations among all the activities that compose the project. An activity that produces an output that is another activity's input must precede the latter activity. Thus, the tip or head of an early activity leads to the tail of the latter activity. This is illustrated at Point 9 in Figure 3.3a, for example, where the tip of the arrow representing "prepare schedule control plan" leads to the tail of the arrow representing "reconcile control plans." Here the output of the earlier activity is a schedule control plan, which is an essential ingredient for reconciling all control plans.

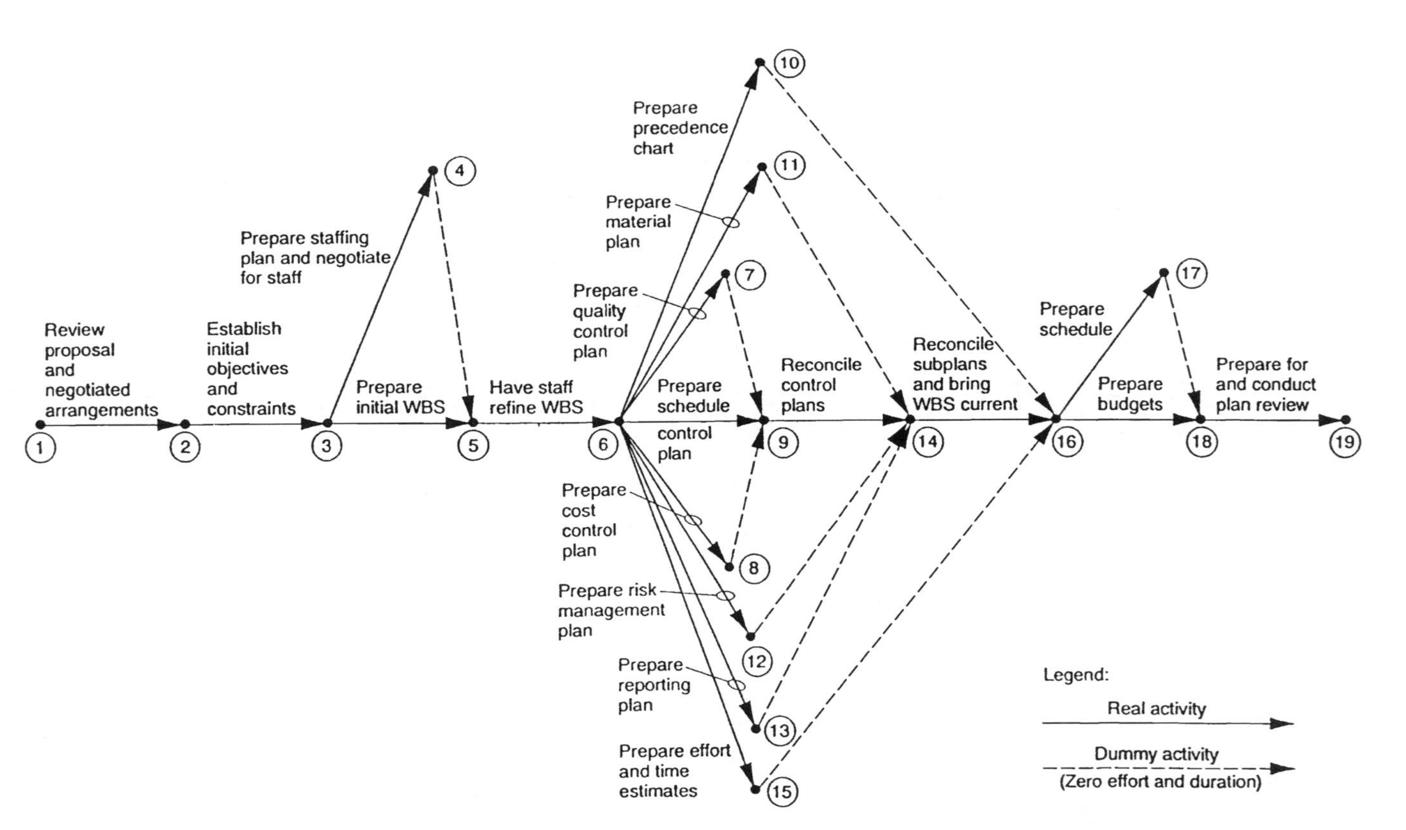

Figure 3.3a. The planning segment of a precedence chart for the development of a spacecraft attitude and articulation control subsystem.

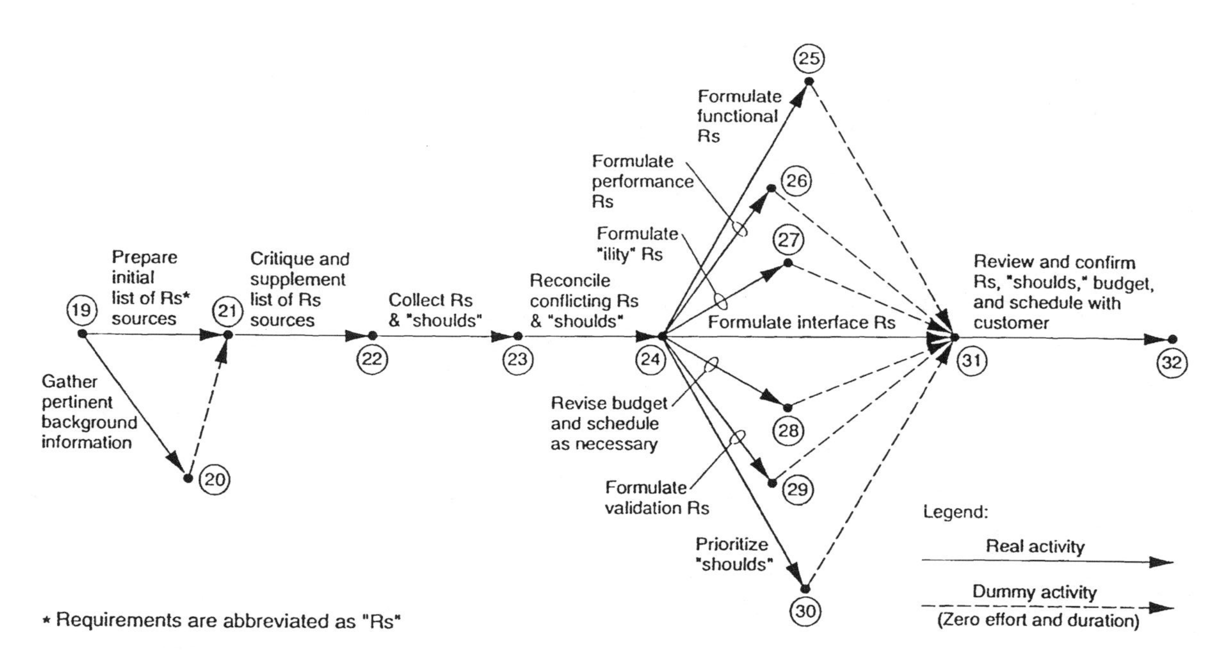

Figure 3.3b. The requirements segment of a precedence chart for the development of a spacecraft attitude and articulation control subsystem.

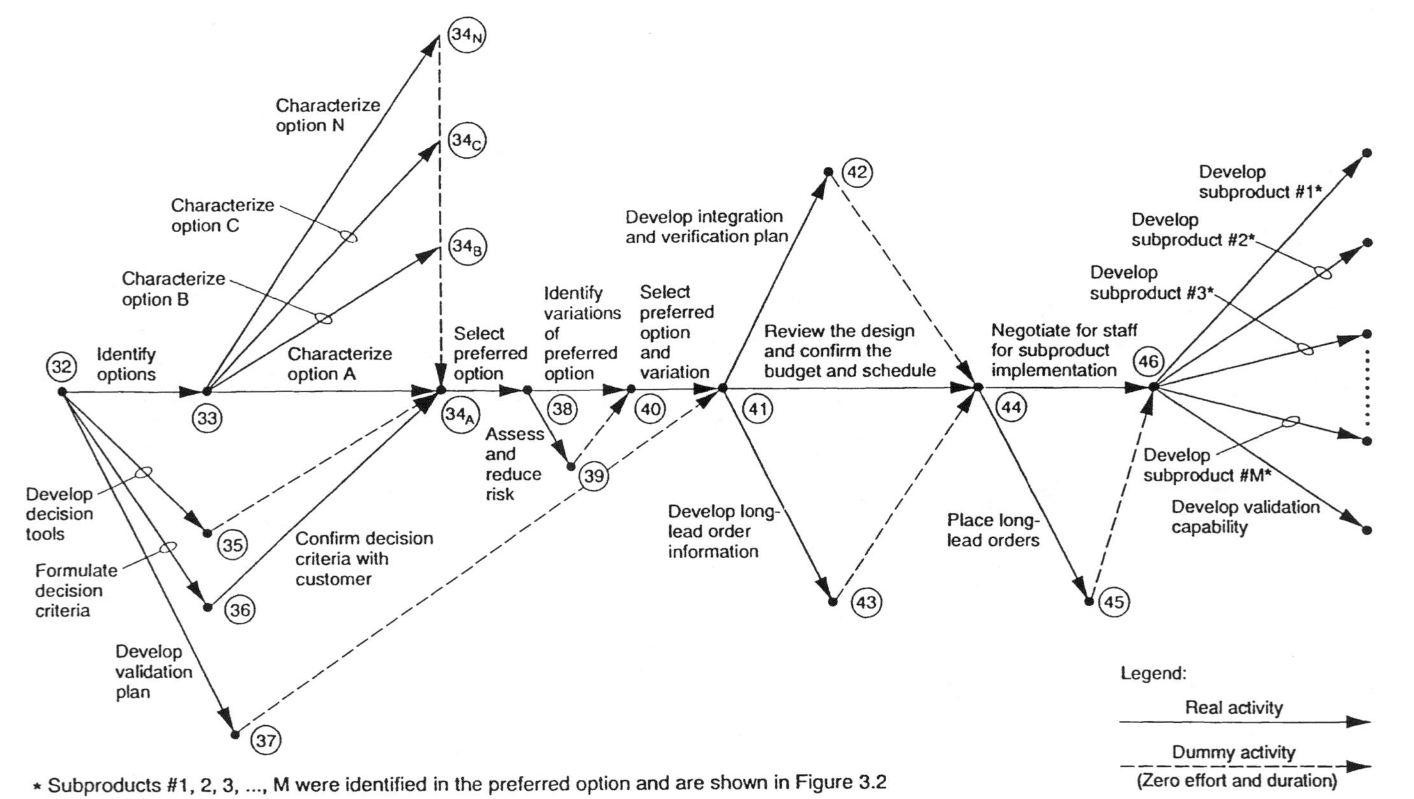

* Subproducts #1, 2, 3, ..., M were identified in the preferred option and are shown in Figure 3.2

Figure 3.3c. The design segment of a precedence chart for the development of a spacecraft attitude and articulation control subsystem.

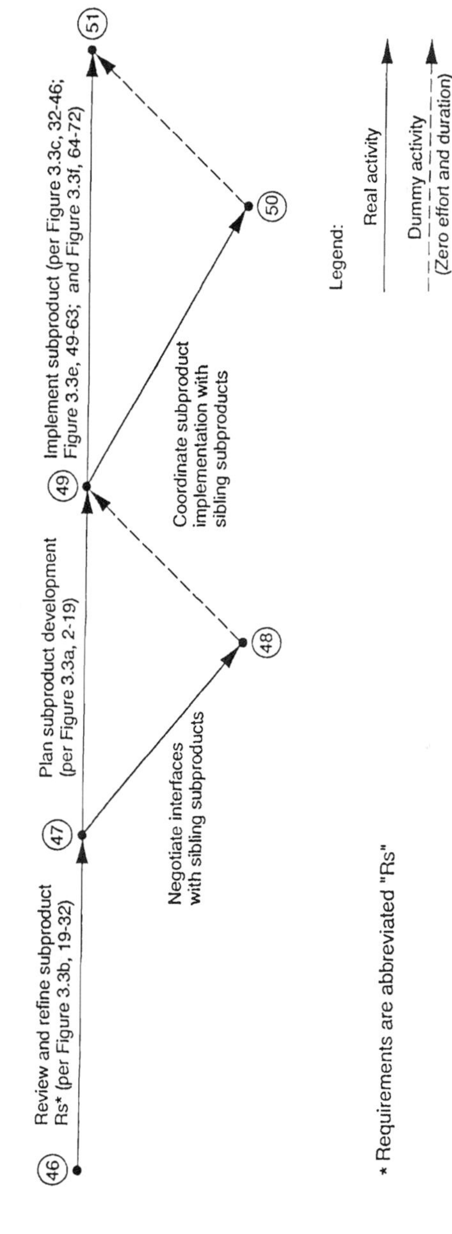

Figure 3.3d. The sub-subsystem implementation segment of a precedence chart for the development of a spacecraft attitude and articulation control subsystem.

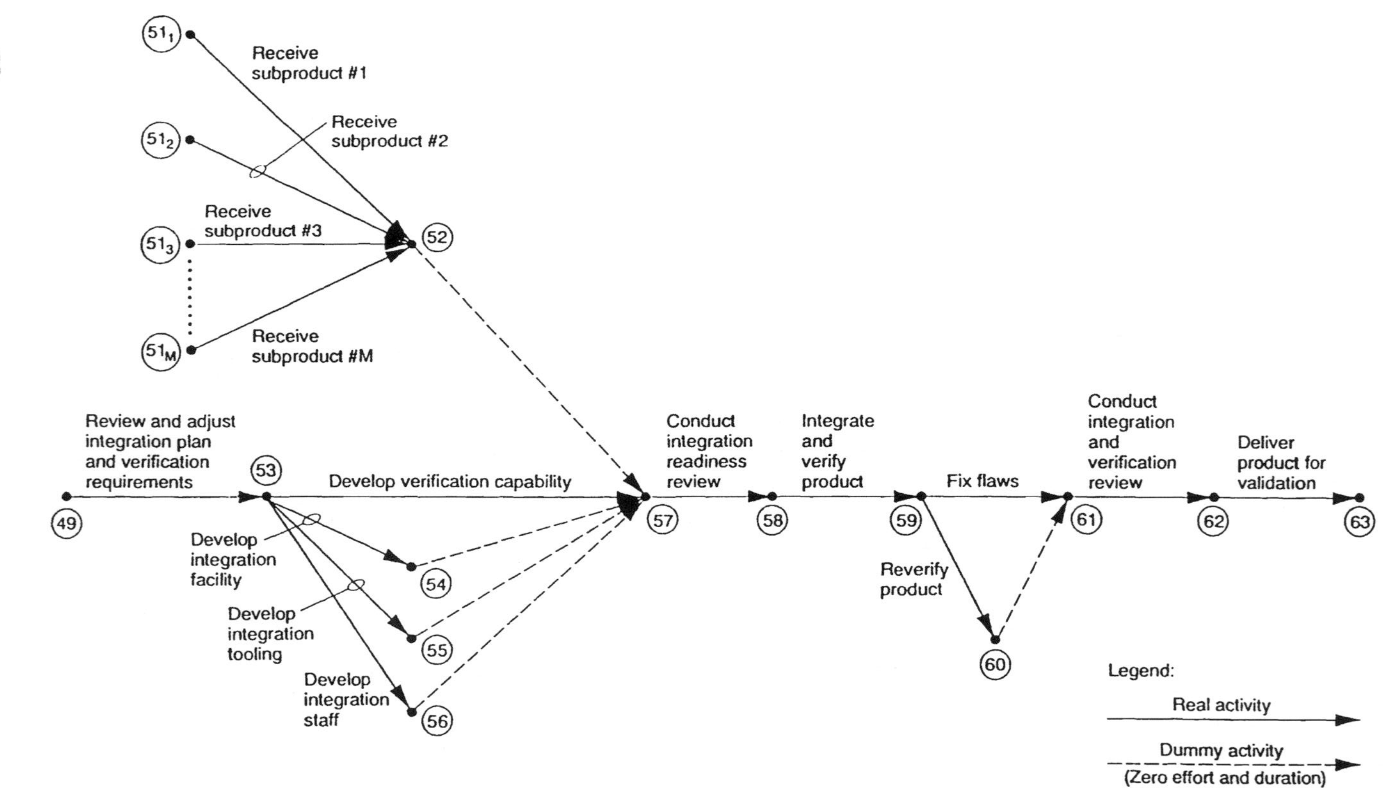

Figure 3.3e. The integration segment of a precedence chart for the development of a spacecraft attitude and articulation control subsystem.

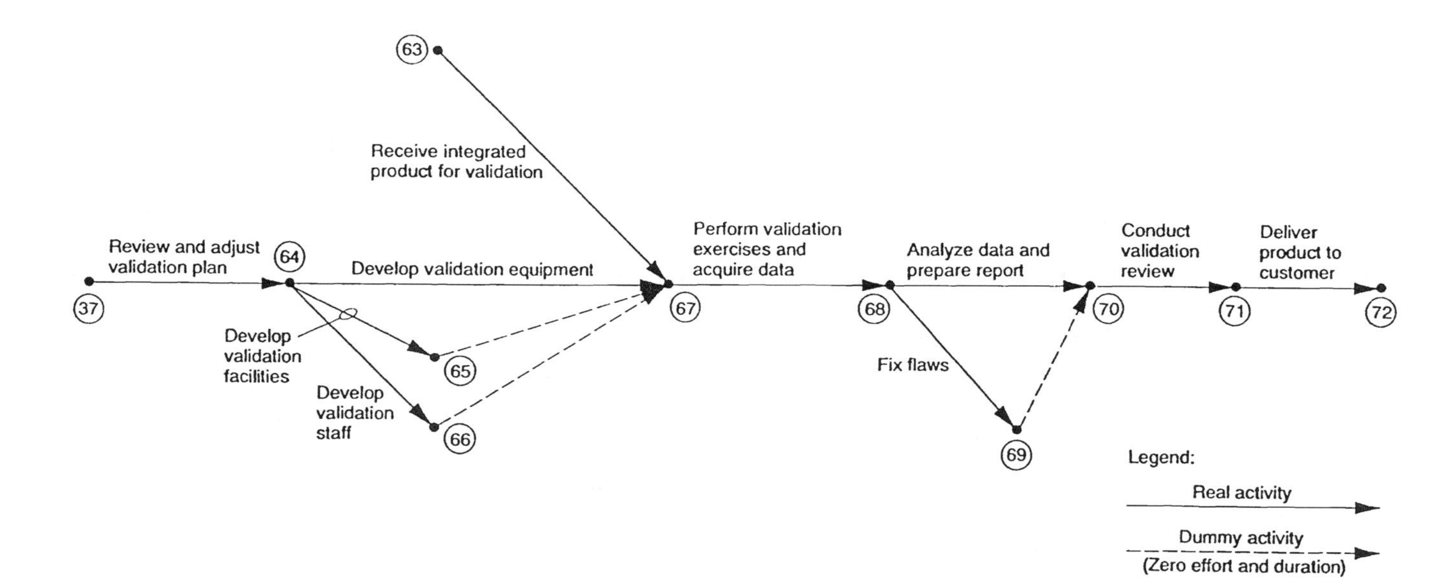

Figure 3.3f. The validation segment of a precedence chart for the development of a spacecraft attitude and articulation control subsystem.

Sometimes the products of two or more activities must be available before a subsequent activity has all the inputs that it needs to start. This is illustrated at Point 67 in Figure 3.3f, where validation equipment, validation facilities, and validation staff, as well as the integrated AACS, must all be available before the integrated AACS can be validated.

Similarly, the conclusion of a given activity can sometimes enable several subsequent activities. This is illustrated in Figure 3.3b at Point 24. Here, seven activities all use as inputs the output of the preceding activity, "reconcile conflicting requirements and 'shoulds.'"

Figures 3.3a–f also contain some dummy activities, shown by dashed lines. Dummy activities have zero effort and duration and are included to keep the scheduling algorithm clear. If two parallel activities had the same nodes for their beginnings and ends, then the algorithm could not distinguish between them. To avoid this dilemma, each of the parallel activities is given a different ending node and then connected to the common ending node by the dummy activity. This is illustrated in Figure 3.3c where the activity "negotiate for staff for subproduct implementation" and the activity "place long-lead orders" occur in parallel. The former activity is labeled Activity 44–46 and the second is labeled Activity 44–45. A dummy activity, Activity 45–46, is used to show that both Activity 44–46 and Activity 44–45 must be completed before activities whose tails lie at Point 46 can be started.

We will return to the subject of precedence charts when we discuss estimating and network techniques of scheduling and budgeting.

IV. SUBPLANS: A CHECKLIST OF MAJOR PLAN ELEMENTS

When thinking of a plan, it is natural to think mainly in terms of performance or work accomplishment. This is indeed the central part of any plan, but it is seldom the whole plan. The entire list of subplans is

1. Performance or accomplishment plans
2. Staffing plans
3. Quality control plans
4. Equipment and material plans
5. Work authorization plans
6. Cost control plans
7. Schedule control plans
8. Reporting plans
9. Risk plans

Subplans 2 through 9 are discussed in turn. All subplans that apply should be included in the project plan.

A. Staffing Plans

The staffing plan tells who will be responsible for each part of the WBS. This plan is easily represented by attaching a personal name to each element in the WBS, thereby creating the basic project organization chart. Basic charts may be augmented to include staff positions for selected basic positions, for example, an administrator or secretary as staff to the project manager.

The project organization chart shows reporting relationships *for the project* and should not be confused with the company's or department's organization chart. Also, the project organization chart is not a status chart. Thus a senior individual who is responsible for a detailed portion of the project may report *on the project* to a junior person who integrates the senior person's work with the work of others.

Nor is the project organization chart a device for aggregating the work to be done by individuals. Thus, the chart should *not* gather together different responsibilities for a single individual in order to have the individual's name appear just once on the chart. To do so would obliterate the project's hierarchical relationships, which are exactly what it is intended to show.

While project organization charts show project-reporting relationships, they do not represent all project communication lines. Indeed, well-functioning projects have a spiderweb of such lines. However, project organization charts do identify the arbitrators of any conflicts that may arise: the individual who occupies the lowest-lying project organization chart node that spans the conflicting parties is the arbitrator.

Significant negotiations are often required to obtain commitments of personnel necessary to fulfill the staffing plan. The project manager can prepare for these negotiations by sketching out a tentative plan, including estimates of the efforts required of each key participant. Prospective project members should make these estimates whenever possible.

Some project managers try to economize on the effort needed to staff their projects by assigning individuals to two hierarchically related WBS elements. This arrangement, however, should be avoided if possible, for two reasons. First, individuals either tend to become engrossed in the details of their lower-level WBS element and slight their coordination responsibilities at the higher level, or they tend to focus on their higher-level responsibilities and slight their lower-level ones. When coordination responsibilities are slighted, interface problems are likely to be ignored or neglected when they are small and easy to resolve, which allows them to grow until they are larger and more difficult to resolve. When lower-level responsibilities are slighted, their neglect is likely to be unnoticed or unattended because individuals are in effect supervising themselves. In either case, slippage and inefficiencies are likely to result instead of economies.

The second reason an individual should not manage WBS elements at two levels is that doing so creates difficulty for other lower-level managers. Since the manager of a higher-level WBS element arbitrates conflicts among the managers at the next lower level, an individual responsible for two levels is both advocate of a lower-level element and arbitrator of conflicts at that level. What should another individual at the lower level make of this duality? Is the two-level individual a peer or a boss? Can the two-level individual be both a strong advocate for the lower-level element and an objective arbitrator? These questions are so vexing that two levels of responsibility for a given individual should be avoided, if possible.

To avoid two-level assignments, individuals can be assigned to two different projects (or subprojects) instead, perhaps in a highly technical capacity on one and in a more managerial capacity on the other. While such assignments require interproject planning, they can be arranged when the parent organization has concurrent projects and attention is paid across the organization to avoiding two-level assignments.

Interproject staff planning will not only minimize two-level assignment problems, it will help optimize staffing situations overall, and it is strongly recommended. In addition, it will promote cross-fertilization and thereby minimize redundant efforts.

The decisions and negotiations that are needed to complete a staffing plan are significant, and the project's overall budget and schedule should provide for them accordingly.

B. Quality Control Plans

The quality control plan contains the scheme for assuring that the project will produce a good product. Thus, it tells who is going to check what, when the checking will be done, and what time and resources are required. It may range from a pair of trained eyes that check the product just before it goes out the door to an elaborate and thorough check and cross check of everything done.

The quality control plan should be well known to the entire project staff. If one person is to be responsible for assuring quality, that person should prepare the plan, consulting appropriately with those who will do the work that will be checked and with the checkers, as well as with the project manager, so that no one will be surprised.

C. Equipment and Material Plans

The equipment and material plan pertains to the physical resources needed to accomplish the project. It begins with a list of items needed, the dates that they are needed, suggested sources of supply, and the lead times necessary to obtain the items, including time to obtain price quotations, shipping time, and time to clear customs if applicable. This information, together with knowledge of the organization's procurement practices, is then used to establish the sequence of activities and the milestones for specifying and ordering every piece of equipment, material, and any necessary physical installations, including plumbing and electrical hookups, needed on the project.

Task leaders should prepare equipment and material plans for their respective parts of the overall plan. The project manager should check these plans to be sure that overall project objectives, particularly schedule objectives, will still be met. Serious slippages commonly occur because of inadequate provision for equipment and material leadtimes.

D. Work Authorization Plans

The work authorization plan is the project manager's scheme for approving successive stages of work. It consists of periodic reviews and evaluations to establish the readiness and appropriateness of each task to proceed and a means of authorizing each task leader to proceed when appropriate.

A primary virtue of work authorization plans is that they enable the project manager to revise individual task plans in order to reallocate resources among tasks as overall project needs change. While a project manager can conceivably retrieve resources that were once assigned to task leaders, it is psychologically difficult to do. An easier approach is to release resources incrementally from time to time. This way, the task leaders are less likely to feel abused when they are required to revise their individual plans as a means of optimizing attainment of overall project objectives.

Sometimes the project manager's customer also authorizes work in increments which are then typically called *phases*. The customer's motivation is the same, namely to be able to examine the results of early work before setting the course and approving later work.

E. Cost Control Plans

Cost control is based upon cost expectations (i.e., task budgets), cost measurements, budget-measurement comparisons, and revisions of plans and budgets to achieve budget objectives when discrepancies are detected. The cost control plan therefore specifies what budgetary details are needed, what costs will be measured, and what comparisons will be made. It also specifies what techniques will be used to collect and process the information and review it in a timely way so that suitable corrective actions may be taken. These activities require resources and time that must be provided in the overall project plan.

Cost control is discussed in detail in Chapter 4.

F. Schedule Control Plans

Schedule control is based on expectations (i.e., performance schedules), measurements of performance versus time, expectation-measurement comparisons, and revisions of performance plans as needed to achieve performance and schedule objectives. Thus they resemble cost control plans in form. The schedule control plan accordingly specifies what performance details will be monitored, when they will be monitored, and by whom. Provisions for these activities must be included in the overall project plan.

Schedule control is discussed in detail in Chapter 4.

G. Reporting Plans

Every project manager will want to know what is going on—all the time. And the customer will want to know on a periodic basis. Thus every project needs a reporting plan that identifies who reports to whom, what is reported, how often reports are made, and how widely the information is distributed.

In order to make overall project reporting as efficient as possible, internal reports (i.e., for those who work on the project) should be coordinated with reports to the customer. Information should be detailed and organized to serve multiple users without recalculation or reformatting whenever possible. Since satisfying customer needs is paramount, their reporting needs should be determined before establishing internal reporting requirements.

As in the case of the other subplans, time and resources must be provided for reporting in the overall project plan. Specific reporting techniques are discussed in Chapter 8.

H. Risk Plans

To plan for risk is to consciously search out potentially important adverse occurrences and then determine how they can be (1) accommodated as is if they should happen, (2) first reduced and then acommodated, or (3) eliminated or avoided altogether. Thus, risk planning goes beyond merely having overall contingency allowances in case something goes wrong. Rather, risk planning involves detailed consideration of possible adverse occurrences and an action plan that promises the best cost/benefit results given the resources available.

Risk planning is also known as risk management and is discussed in Chapter 5.

V. ESTIMATING TECHNIQUES

The time and expense of performing a project can be estimated in several ways, which are described in this section. No one of the ways is absolutely foolproof, and prudent project managers will use as many as possible and compare their totals. If significant differences are found, then each estimate should be examined to try to account for the differences. Generally this examination will reveal important assumptions or oversights in one or more of the estimates which can then be properly acknowledged, thereby reconciling the differences.

A. Top-Down Estimating

One estimating technique is called *top-down estimating*. In this approach, large blocks of effort are estimated by comparing these blocks to those with which one has had experience. The large blocks are typically the major pieces of work identified in the work breakdown structure. The estimates for the large pieces are then added together to obtain estimates for the entire project.

The top-down approach naturally works best when the blocks in the current project closely resemble those in earlier projects and when accurate records of time and expense exist for the earlier efforts. It works hardly at all when one has no earlier experience to refer to. Its greatest virtues are that it is quickly applied and that it provides for necessary interactions, even if they are obscure.

B. Bottom-Up Estimating

An alternative to top-down estimating is the bottom-up approach. In this technique, small tasks are estimated individually and then summed to arrive at the total project time and expense. The small tasks are typically the items in the precedence chart, which is described in Section III of this chapter.

The bottom-up approach depends upon being able to gauge how much time and other resources are needed to perform very elemental tasks. Often one can estimate small increments or elements quite well even when overall efforts cannot be estimated very satisfactorily. This is particularly true when many specialties are involved and no one person has enough breadth to be able to integrate them all together to produce a top-down estimate.

A possible problem with bottom-up estimating is that the final overall estimate may contain too much contingency allowance for the situation. This arises when each individual estimator provides a minimum contingency allowance to his or her part. In the aggregate, these allowances total more than warranted by the collection of individual elements taken as a whole. Another problem is that this approach can be quite time-consuming, for it requires each individual task to be enumerated and characterized.

C. Standard Costs and Times

An adjunct to either top-down or bottom-up estimating is the use of standard times and costs. Certain tasks in some organizations (e.g., drilling holes, installing wire, pouring concrete, curing adhesives, taking samples, analyzing samples, writing computer code, etc.) are virtually standardized operations. The time and expense involved in such operations are well known and do not vary appreciably from one project to another. Whatever variances exist are randomly distributed and do not depend upon the particular project.

Standard times and costs can be used in preparing estimates without having detailed insight into the tasks themselves. This approach can be used of course at whatever level the standards apply, that is, either large work blocks or small elemental tasks, but standard information is most likely to exist for individual elemental tasks.

D. Historical Relationships

Other useful adjuncts to top-down and bottom-up estimating are the historical relationships that may exist among different parts of similar projects. The efforts spent on documentation, quality control, accounting, supervision, and other indirect functions often have rather stable ratios to the expenses for direct labor. Similarly, the costs for piping, instrumentation, insurance, architectural services, etc. may have stable relationships in certain circumstances with the costs of major pieces of equipment such as boilers, heat exchangers, fractionating columns, and so forth.

A special version of the concept of historical relationships is the empirical fact that equipment costs typically vary as a function of capacity or size by a power law of the form

$$\frac{\text{Cost}_2}{\text{Cost}_1} = \left(\frac{\text{Capacity}_2}{\text{Capacity}_1}\right)^n$$

where the value of n varies with economies or diseconomies of scale. If n is 1.0, then neither economies nor diseconomies exist. If n is greater than 1.0, then diseconomies exist, for example, in the case of large-diameter front-surface optical mirrors. If n is less than 1.0, then economies exist, for example, in the case of microwave antennas that operate at a given frequency. Values of n when it is less than 1.0 are commonly in the range 0.6 to 0.7.

The power law is useful in extrapolating data from one or two cases where costs are known or can be easily obtained (e.g., from recent quotations or bids) to cases where they must be estimated. It assumes, of course, that there are neither physical limitations to making the item being estimated nor any step-function changes in the design or manufacturing approach between the item whose cost is known and the item whose cost is being estimated.

Manufacturing costs may also vary systematically, but in a complex way, with such parameters as number of separate parts, type of material, amount of material removed by machining, weight or volume of the finished article, and the number of units produced to date. If sufficient historical data exist, statistical techniques can be used to determine parametric relationships of cost versus the variables listed. These relationships can then be used to predict the cost of new items that are made essentially the same way as those for which data exist.

E. Simpson's Rule

The actual time or cost of a piece of work may be different from nominal or most likely time or cost. Also, the actual time and actual cost are unlikely to exceed some maximum value and unlikely (or unable) to be less than some minimum value. These facts give rise to an estimating technique called Simpson's Rule, which uses nominal, minimum, and maximum values to calculate the expected value of a parameter:

$$X_{exp} = \frac{X_{min} + 4X_{nom} + X_{max}}{6}$$

where

X_{exp} = the expected value of x
X_{min} = the minimum possible value of x
X_{nom} = the nominal or most likely value of x
X_{max} = the maximum value of x

Simpson's Rule can be used to calculate expected times and costs for each activity in the precedence chart. As described in Section VI, the expected and nominal times for the activities can be used together to determine the project's overall duration, due dates for individual activities, and the project's overall schedule contingency allowance. And as discussed in Section VII, the expected and nominal costs for the activities can be used together to determine the project's overall budget, budget allocations for individual activities, and the project's overall budgetary contingency allowance.

The use of Simpson's Rule can reduce resistance to estimating time and cost. Project team members often resist estimating because they fear that their numbers will be wrong but will become commitments nevertheless. This typically occurs when single values of time and cost are sought for each activity. Using Simpson's Rule tells team members that they are not being held to single-value estimates and removes much of the resistance to estimating.

F. Documenting Estimates

As will be seen in Chapter 4, the estimates used to prepare the project schedule and budget will provide the basis for controlling the project. It pays, therefore, for the project manager to ensure that all the assumptions surrounding the estimates are documented. Otherwise, it will be impossible to tell whether the schedules and budgets based on the estimates are still valid if actual performance does not correspond to the schedules and budgets.

Also, astute project managers, and indeed astute organizations, recognize the value of saving both time and cost estimates and actual time and cost data for future reference. In the final analysis, no better way exists to estimate the next project than to use data and insight developed on related or similar previous projects.

VI. SCHEDULING TECHNIQUES

The essence of project scheduling is to apply time estimates to the precedence chart and then make various adjustments (which are discussed later) so that the entire project can be accomplished within the time and budget allowed.

Bottom-up estimates, if available, apply to individual activities in the precedence chart. This is illustrated in Figure 3.4, which is based on Figure 3.3c from Point 32 to Point 46. Top down estimates apply to larger segments of the chart, each of which subsumes many individual activities. This is illustrated in Figure 3.5, which is also based on Figure 3.3c from Point 32 to Point 46 but without detailed (bottom-up) estimates for the portion from Point 32 to Point 41.

When bottom-up estimates are used, one or more necessary coordinating activities may be overlooked in the estimated segments. The time required for these activities must also be estimated and applied to the precedence chart.

Once the time estimates have been applied to the precedence chart, the longest sequence from project beginning to project completion can be determined. The procedure can be illustrated by referring to Figure 3.4 and the following calculations.

First, the earliest possible starting time and earliest possible completion time are calculated for each activity. No activity can start sooner than the completion of all activities that must precede it. When two (or more) activities immediately precede a given activity, both (or all) of them must be completed before the given activity can begin. Thus, completion of the last parallel activity governs the start of subsequent activities. Earliest possible starting and completion times are shown in Columns 3 and 4, respectively, of Table 3.1, which relates to Figure 3.4.

The next step is to calculate the latest permissible starting times for each activity. These times are the latest times that activities can be started without causing any subsequent activity to be so late that it delays the last task in the project. The latest permissible starting times are calculated, of course, by starting at end of the last activity and working backward. These times and the corresponding latest permis-

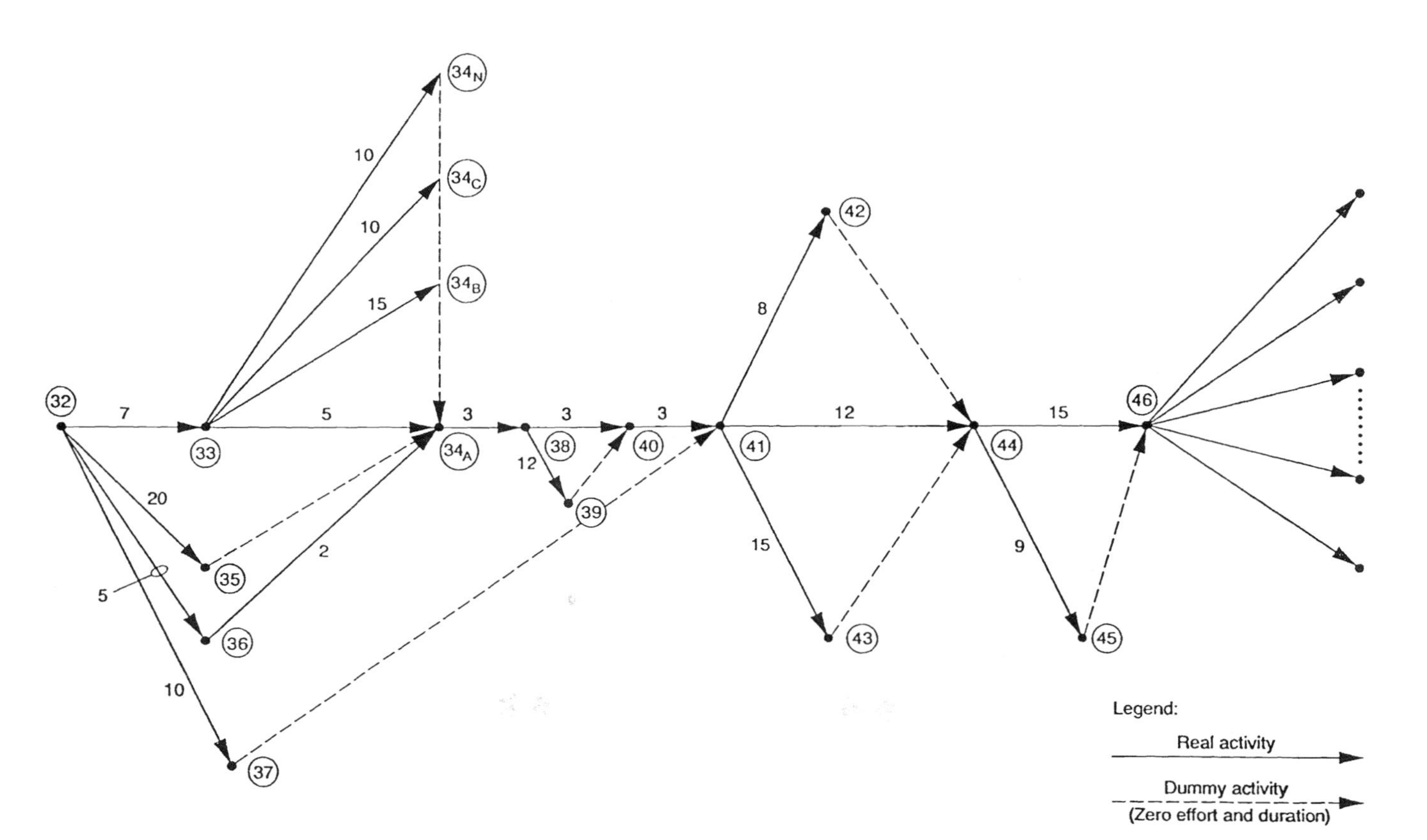

Figure 3.4. Time estimates for the precedence chart shown in Figure 3.3c, using Option B.

Table 3.1. Critical Path Calculations for the Precedence Diagram Shown in Figures 3.3c and 3.4, Using Option B

Task	Duration	Earliest possible start time	Earliest possible completion time	Latest permissible start time	Latest permissible completion time	Slack	Critical task
32-33	7	0	7	0	7	0	Yes
32-35	20	0	20	2	22	2	No
32-36	5	0	5	15	20	15	No
32-37	10	0	10	30	40	30	No
$33\text{-}34_B$	15	7	22	7	22	0	Yes
$35\text{-}34_B$	0	20	20	22	22	2	No
$36\text{-}34_B$	2	5	7	20	22	15	No
37-41	0	10	10	40	40	30	No
$34_B\text{-}38$	3	22	25	22	25	0	Yes

38-39	12	25	37	25	37	0	Yes
38-40	3	25	28	34	37	9	No
39-40	0	37	37	37	37	0	Yes
40-41	3	37	40	37	40	0	Yes
41-42	8	40	48	47	55	7	No
41-43	15	40	55	40	55	0	Yes
41-44	12	40	52	43	55	3	No
42-44	0	48	48	55	55	7	No
43-44	0	55	55	55	55	0	Yes
44-45	9	55	64	61	70	6	No
44–46	15	55	70	55	70	0	Yes
45-46	0	64	64	70	70	6	No
46-46′[a]	0	70	70	70	70	0	Yes

[a]The dummy activity 46-46′ is added to allow the algorithm to run its course and identify the latest of the earliest possible completion times of tasks that end at 46.

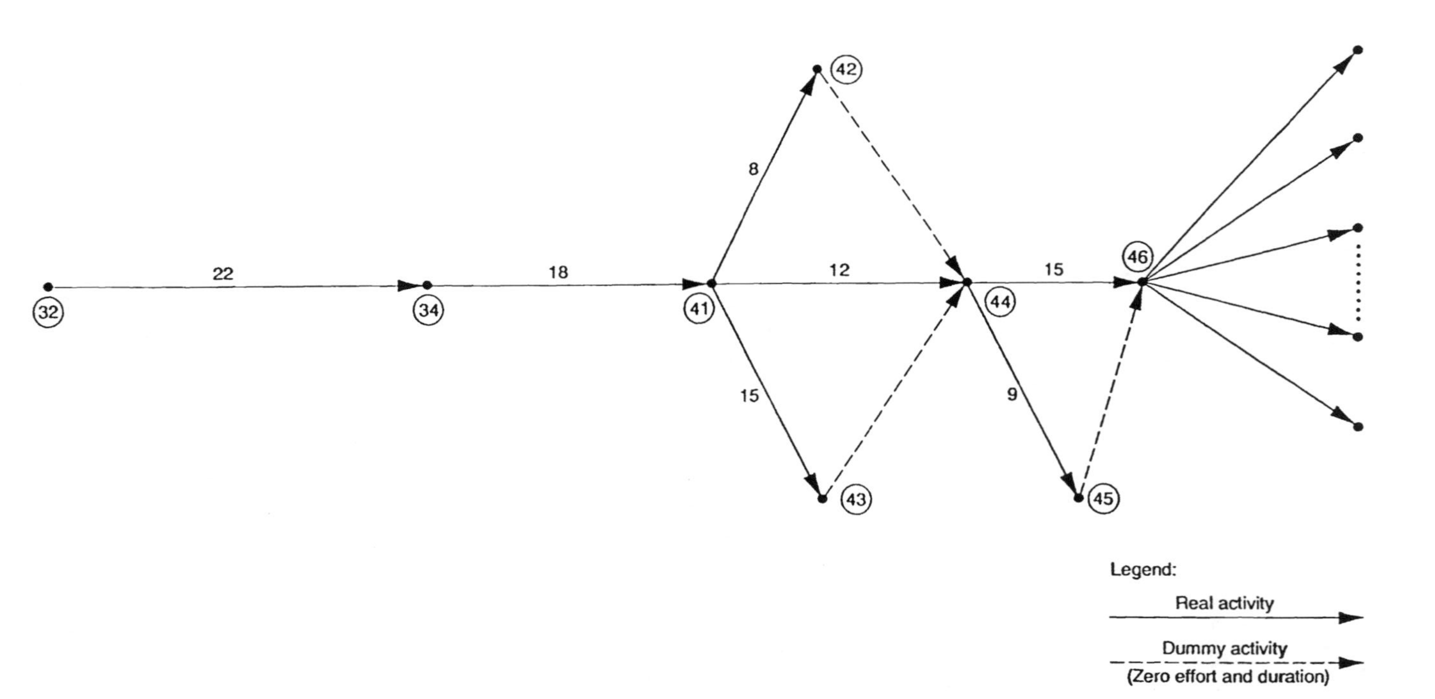

Figure 3.5. A simplified version of Figure 3.4.

sible completion times are shown in Columns 5 and 6, respectively, of Table 3.1.

The difference between the earliest possible starting time and latest permissible starting time for each activity is called the *slack* for the activity. Slack refers to the amount of delay that an activity could have, relative to its earliest possible starting time, without causing the last activity to be late. Activities that have zero slack are said to be *critical*. Critical activities cannot be delayed without causing the overall project to be late. The critical activities form a sequence that is called the project's *critical path*. The critical path for the example in Figures 3.3c and 3.4 is shown in Figure 3.6. Figure 3.7 is a time-scaled version of this critical path.

If the overall project duration, that is, the duration of the critical path, is within acceptable bounds, no adjustments in the project schedule are needed. In this case, the project manager must simply make sure that (1) no critical activity slips its schedule to the point that it causes an unacceptable slip in the overall project, taking contingency allowances into account and (2) no activity that is not critical slips so much that it becomes critical.

If the overall project duration as just estimated is unacceptably long, however, then the project manager must find ways to shorten the project duration. The manager may also wish to shorten the duration of a segment of the project if it involves an expensive item whose cost could be decreased by using it more intensively or by minimizing delays between its successive uses.

One approach to shortening a project's duration is to double up on one or more of the critical activities so that the total project may be done in less time. In some cases, this can be done without extra cost or at least no more cost than that saved by doubling up. In other cases, extra funds may be required (e.g., to pay for overtime or for extra coordination required when an activity is subdivided and as-

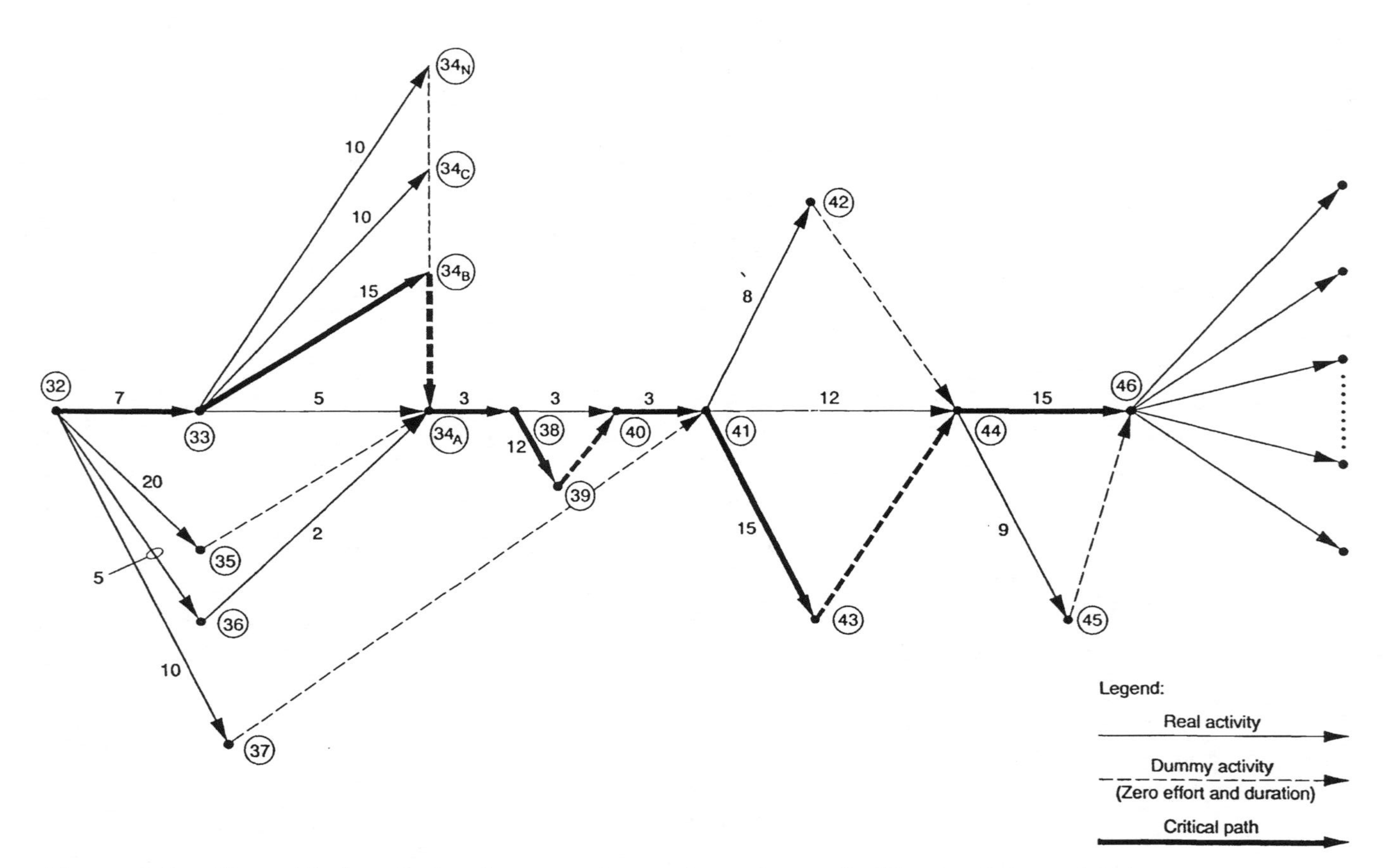

Figure 3.6. The critical path for the precedence chart shown in Figures 3.3c and 3.4.

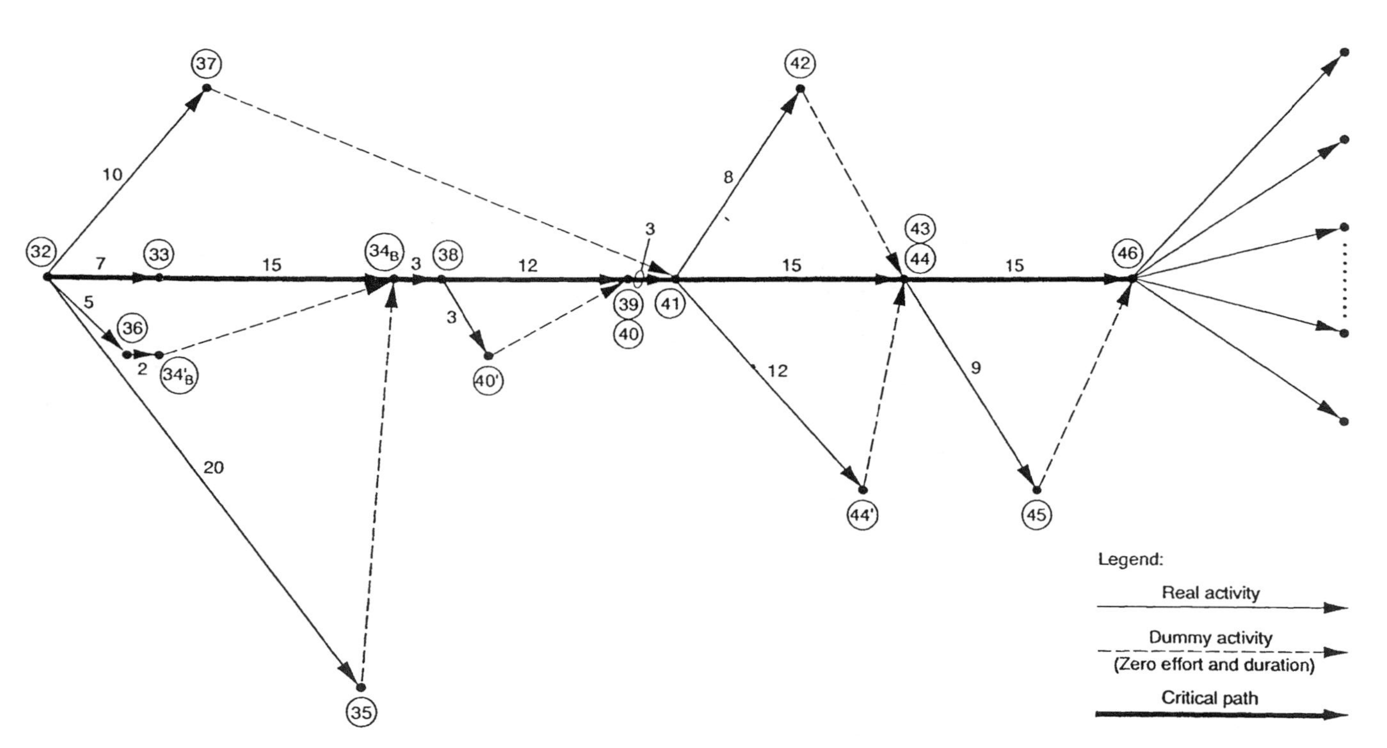

Figure 3.7. A time-scaled version of the critical path shown in Figure 3.6.

signed to different individuals). If extra costs are involved, they must be agreed to by the customer before they are incurred. In still other cases, it may not be possible to double up—two elephants, each pregnant for 10½ months, are not equivalent to one elephant pregnant for 21 months! Another approach is to negotiate a reduction in the total project scope of work so that it can be accomplished within an acceptable time. And of course, sometimes an extension in the time allowed to complete the project can be negotiated.

Quite possibly, a suitable combination of these approaches is best overall. The preferred mix will depend upon the customer's or boss's priorities, and the project manager should be prepared to discuss the pros and cons of various ways to define and schedule the project.

The above approach to scheduling uses what might be called *exact* time estimates for each task. Some refinements are possible if it is recognized that the estimates might not be exact. If, for example, the probability distribution of possible times is known for each task, then Monte Carlo techniques can be used to determine the overall duration of the project.

A particularly useful refinement uses information from Simpson's Rule, discussed in Section V.E of this chapter. First, the overall duration, or "commitment duration," is calculated as above, using the larger of each activity's nominal time or expected time. Next, a "target duration" is calculated as above using the smaller of each activity's nominal time or expected time, with corresponding target due dates for each activity. Finally, the project's schedule contingency allowance is calculated as the difference between the commitment and target durations. This contingency allowance is a valuable part of risk management, which is discussed further in Chapter 5.

This scheduling approach is greatly aided by using a computer-based project management information system

(PMIS) to perform the calculations, which are admittedly tedious and subject to error. Commercially available PMISs will determine overall project duration, critical paths, and slacks from time estimates and precedence relations. These systems will also graph the schedule as well as the WBS and precedence chart.

Another reason for using a computer-based PMIS in scheduling is that it facilitates keeping the schedule current when more accurate estimates become available. Typically, the quality of time estimates for later activities can be improved as the project progresses. By using a computer to keep track of the schedule, it can be easily revised when better time estimates are available.

The scheduling approach just described is variously known as the critical path method (CPM) or as program evaluation and review technique (PERT). The particular representation used here is called *activity on arrow* because the activities are in fact represented by the arrows or lines of the drawing. An alternative representation is *activity on node,* where the activities are shown by points, boxes, or circles connected by lines. In the case of activity on arrow, the drawing is time-scaled. In the case of activity on node, no time scale is implied by the horizontal axis. Many computer printouts use the activity on node representation because it saves paper.

VII. BUDGETING TECHNIQUES

Budgeting refers to allocating financial and other resources to the elements of the work breakdown structure. If all resources were available in unlimited amounts, then scheduling alone as discussed in Section VI of this chapter would provide the information needed to assign resources to various parts of the project. Rarely, however, are all resources so abundant. Thus, project managers must decide which resources and how much shall be applied to different activities in order to satisfy project objectives.

The more capable computer-based PMIS will assign personal and other resources to activities according to the resources' capabilities, availabilities, and unit costs. A resource's capability is given in terms of the activities that it can perform. A resource's availability is given in terms of specific dates or hours that it can serve the project. The PMIS will optimize the assignment of resources according to the project manager's criteria, e.g., shortest overall duration, lowest overall cost, minimum use of a specific resource, minimum number of a particular type of resource required, etc., or some prioritized combination of these.

Trying alternative resource arrangements is likely to be as fruitful as trying alternative schedule logics in terms of identifying the best project plan. In fact, some alternative schedule logics can be considered only by introducing alternative resource arrangements. Thus, when project managers do what-if analyses, they should consider a range of resource arrangements as well as a range of schedule logics.

Section VI described the use of Simpson's Rule information to determine commitment duration, target duration, target due dates for individual activities, and the schedule contingency allowance. A parallel exists in setting the commitment budget, target budget, initial budget allocations, and the budgetary contingency allowance. Thus, the larger of each activity's nominal cost or expected cost is used to calculate the project's overall budget, or "commitment budget." Similarly, the smaller of each activity's nominal cost or expected cost is used to calculate the project's target budget and to set initial budget allocations for individual activities. And the difference between the commitment and target budgets is the budgetary contingency allowance, to be used in managing risk (Chapter 5).

VIII. REPRESENTING THE PLAN

A fully developed project plan contains so much detail that many people are unable to comprehend its most important features. Accordingly, a project manager may wish to summarize or capsulize it when talking to individuals who are not interested in its intricacies.

Descriptions of the project's main objectives, main approaches, main management techniques, and so on, can be reduced to just a few pages of narrative. If the description exceeds four or five pages, most people who should read it probably will not do so.

Budget summaries can be prepared by major objectives, major tasks, or major blocks of time. Supporting elements, e.g., quality control, can be included with the major elements or listed separately, depending on the need for simplicity or explicitness.

Schedule information can be represented in a bar chart, as in Figure 3.8. This figure gives the same schedule information as Figure 3.4. While it does show time phasing of the tasks, it does not reveal the essential precedence relationships that connect them.

A special version of the bar chart is the Gantt chart, named after the man who developed it. In a Gantt chart, each task bar is shaded to show its percentage completion as of the date of the chart, as in Figure 3.9. *If* it is assumed that the work of each task should proceed at a uniform rate during the task, then the percentage completion data can be directly compared against the date of the chart to determine which tasks are ahead, behind, and on schedule. However, many tasks do not proceed at a uniform rate, so such comparisons may not be meaningful.

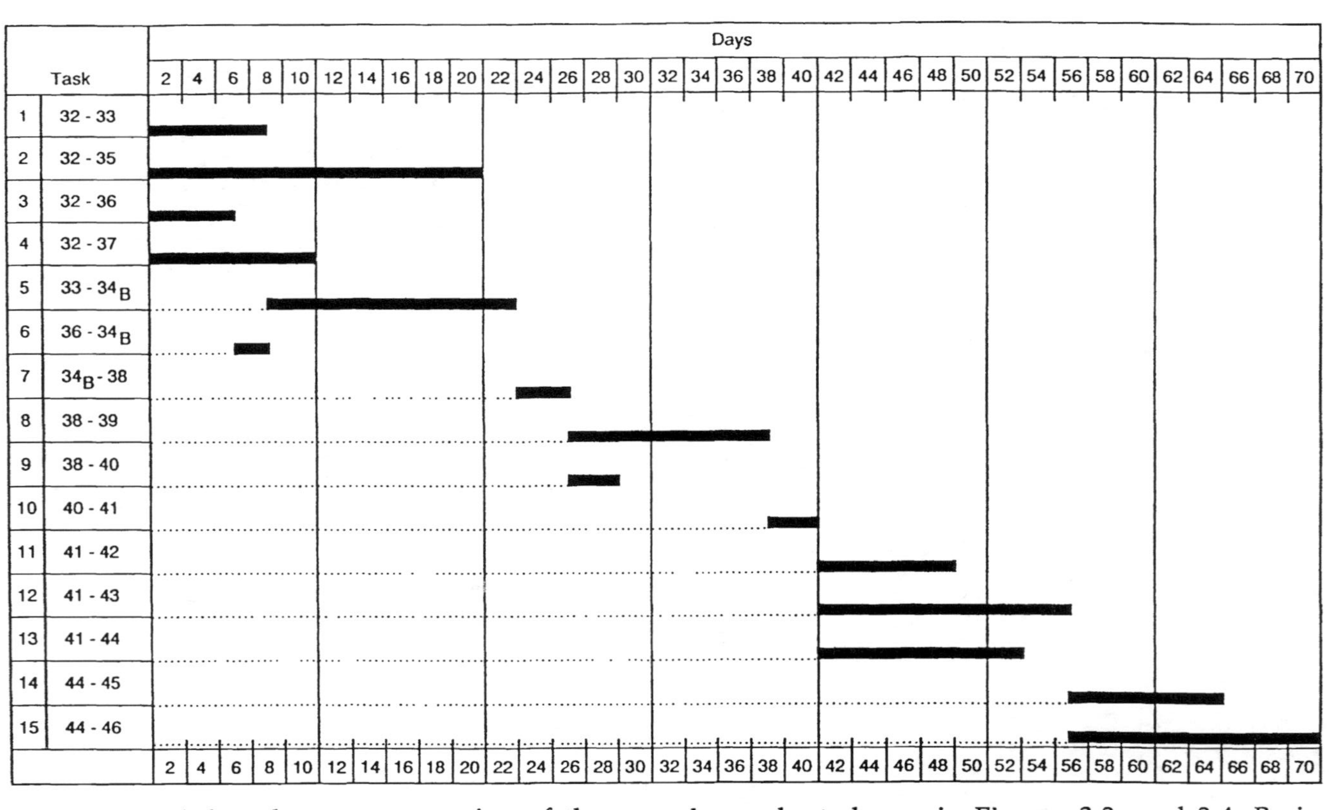

Figure 3.8. A bar chart representation of the precedence chart shown in Figures 3.3c and 3.4. Basis: earliest possible starting times.

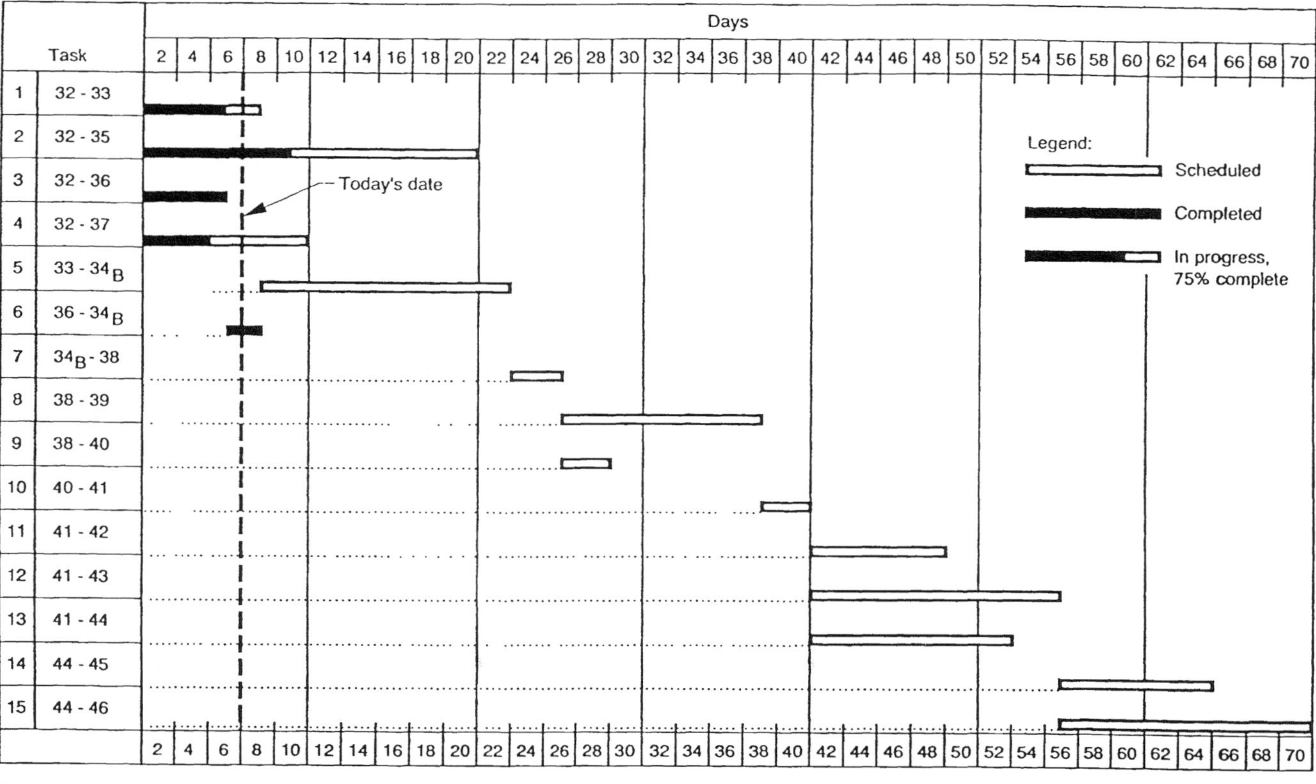

Figure 3.9. A Gantt chart representation of the precedence chart shown in Figures 3.3c and 3.4. Basis: earliest possible starting times.

4

Control Techniques

Project objectives are rarely met on time and within budget unless the project is controlled. By *controlled,* we mean monitored and managed in such a way that deviations from plan are detected and corrected in time for the objectives, indeed, to be met and met on time and within budget. This is the strict notion of control; it is corrective in nature. Closely related, however, is prevention, where early warning signals are used to detect potential deviations even before they occur so that trouble can be averted. Prevention includes both periodic assessments of impending trouble and less frequent, but more probing, assessments called *reviews.*

I. PRINCIPLES OF CONTROL

Control implies

1. Expectations of what should happen
2. Measurements of what has happened
3. Comparisons between expectations and what has happened

4. Timely corrective actions designed to meet the objectives, schedule, or budget

If any one of these four steps is missing, the project cannot be controlled.

This simple yet powerful concept of control contains some important corollaries. First, the expectation or plan must be expressed in terms that are also suitable for measurement; otherwise, it will not be possible to compare measured results with expectations. Second, what is measured should correspond to elements in the plan; otherwise, necessary comparisons cannot be made even if the plan is expressed in measurable terms. And third, the plan and the corresponding measurements must be made at intervals close enough together that when a deviation is detected, enough time, budget, and other contingency allowances exist to correct the situation.

It follows from this last corollary that the less contingency allowance (or slack) available, the more closely a project must be monitored for control purposes. A project manager does not want to find that needed corrections (including, perhaps, completely redoing a segment of the work) cannot be made within the time or resources remaining. Thus, when contingency allowances are minimal, the project plan and the monitoring program must provide information on relatively small increments. Otherwise, the deviations will not necessarily be small enough to be corrected without exceeding available contingency allowances. Conversely, when contingency allowances are generous, the plan and monitoring program can provide information infrequently and on relatively large increments of work, because the allowances can accommodate even relatively large corrections.

The fact that the level of monitoring permissible or required varies as a function of the contingency allowance available leads to the notion that there may be an optimum mix of monitoring and contingency allowance. If an infinite contingency allowance is provided, monitoring can be reduced to virtually nothing: the project manager can wait until the end to see if everything worked out. If it did not, then the work can be redone (for in this extreme case the contingency allowance is infinite). This approach is very cheap in terms of monitoring and infinitely expensive in terms of contingency allowance. While the contingency might not be used, allowing for it in case it is needed can result in scheduling and pricing the project beyond the customer's ability to wait or pay.

At the other extreme, the monitoring program can be both very intensive and extensive, so that incipient deviations from plan are detected and corrected almost instantaneously; this approach uses virtually no contingency allowance. However, while quite cheap in terms of contingency allowance, it is very costly in terms of monitoring. Again, the costly element (i.e., monitoring in this case) may exceed the customer's budget.

An intermediate blend of monitoring and contingency allowance avoids the two very costly extremes and can be suitably effective. This effect is depicted in Figure 4.1, where the costs of monitoring, of contingency allowance, and of their total are shown for a given level of confidence that project objectives, schedule, and budget will be met. This figure shows a minimum in the total cost. At this minimum, the sum is less than the cost in either case where one of the components is zero. That is, a blend of both monitoring and contingency allowance is more cost-effective than either one or the other alone.

Figure 4.2 shows schematically the relation of total costs of control for different levels of confidence that the project objectives, schedule, and budget will be met. Curve A in this figure is the same as the total cost curve in Figure 4.1, while Curve B represents total costs for a higher level of confidence, and Curve C represents total costs for a lower level of confidence.

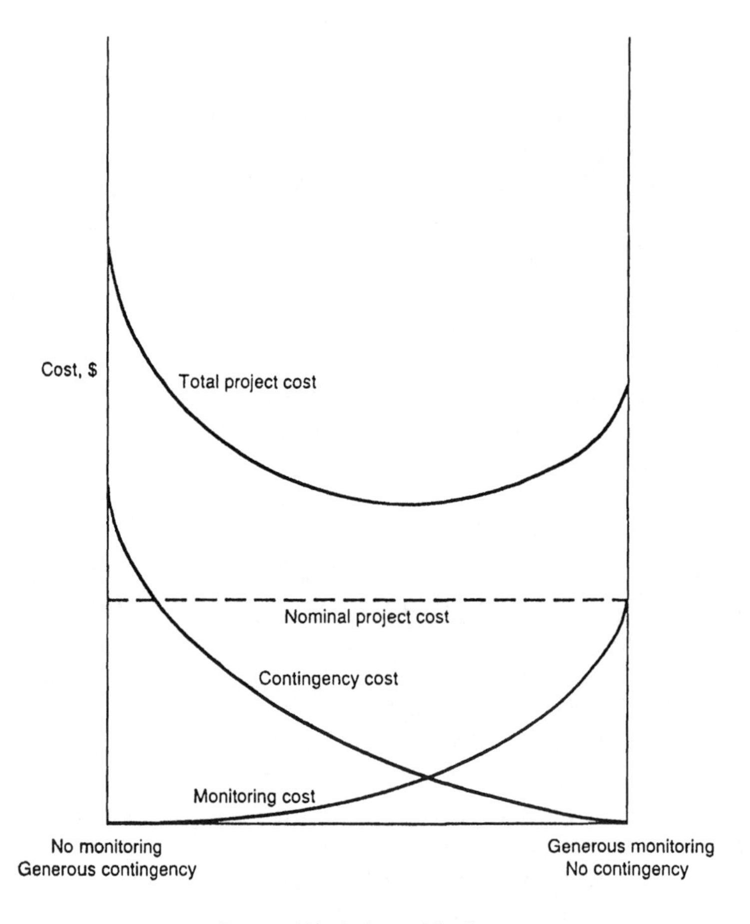

Note: "Nominal project cost" refers to the estimated project cost without provisions for monitoring or contingency.

Figure 4.1. Total project cost versus degree of monitoring and contingency allowance for a given level of confidence (probability) that project objectives will be met.

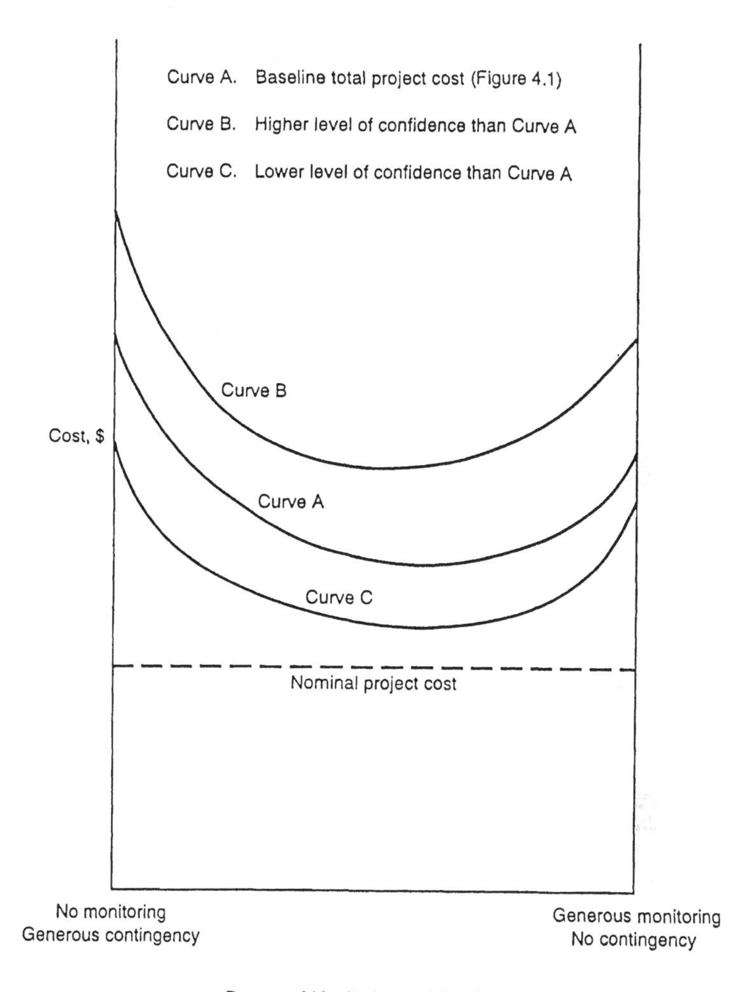

Figure 4.2. Total project cost versus degree of monitoring and contingency allowance for different levels of confidence (probability) that project objectives will be met.

The curves in Figure 4.2 refer only to levels of confidence, or probabilities, that project objectives, schedules, and budgets will be met. In other words, good project control does not guarantee project success; it can only raise the likelihood of success to a very high probability. Even then, an occasional project, appropriately controlled, will not fully meet an objective, schedule, or budget. Nothing in project management is so certain as to prevent all such occurrences.

The general principles just described can be reduced to specific procedures for task or performance control, schedule control, and budget control. They are discussed in the following sections.

II. TASK CONTROL

Task control consists of assuring that the work itself is accomplished according to plan (without regard to schedule or budget, which are handled separately).

A. Detailed Functional Objectives

Task control is based first of all upon detailed functional objectives for the elements in the work breakdown structure (WBS) described in Chapter 3. Each objective should possess the characteristics of a good requirements statement, as described in Chapter 1, so that their accomplishment can be ascertained unambiguously. The increments of work should be no greater than the schedule and budgetary contingency allowances available to fix any mistakes that are found, and preferably much less. Then, if a particular increment of work is found deficient, it can be done within the available allowances and not thwart the attainment of the project's overall objectives, schedule, and budget.

By monitoring in increments that are relatively small compared to whichever of the schedule and budgetary contingency allowances for the segment is more stringent, correction of a single deficiency will not exhaust a major part of the allowance. This will leave some allowance for correcting other increments that might later be found deficient. As a rule of thumb, monitoring increments that are 5 to 10% of available allowances have proven useful. Monitoring increments are discussed further in Appendix D.

B. Quality Inspections

Inspections of work quality are an important aspect of task control and go hand in hand with detailed functional objectives. Each work increment, no matter how simple, needs to be checked *at some level* to assure that it is satisfactory. Project managers dare not find that a trivial item, such as late delivery, wrong specifications, inadequate access, incorrect size, faulty instrument, lack of suitable personnel, and so forth, is precluding a successful project. And, certainly, no major task can go unchecked. Thus, project managers must arrange for timely and appropriate reviews or inspections to confirm either that the work is being done according to plan or that a deficiency exists that must either be corrected or accommodated by changing the plan.

C. Work Orders

Another element in task control is the use of work orders to authorize work increments or packages. By using work orders, the project manager can force personnel working on related parts of the project to coordinate their efforts: they may not proceed until authorized and the authorization will be given only when the project manager is satisfied that their interactions are properly reflected in their respective efforts. This approach minimizes the chance of having to redo a portion of the work because of overlooking interactions of related activities.

Work orders also preserve flexibility for major revisions which may be necessary if contingency allowances are fully consumed. When an original contingency allowance is used up, the only way the project manager can provide for still other possible deficiencies and unforeseen events is to rescope some or all of the remaining work. While it is theoretically possible to cancel work already authorized, it is time-consuming and perhaps costly to do so. On the other hand, if most of the future work is yet to be authorized, some of it can easily be "cancelled," simply by not authorizing it. Thus, using work orders makes it easier for the project manager to rescope future work in light of the total situation and thereby better meet their overall project goals.

III. SCHEDULE CONTROL

Schedule control consists of assuring that the work is accomplished according to the planned timetable. Generally, there is little concern if work is accomplished early, so attention is usually focused on preventing schedule slippage. However, if premature accomplishment would result in cash flow problems or excessive interest charges, project managers should keep major activities from occurring until they are needed.

A. Causes of Schedule Slippage

Schedule slippage is commonplace and should be a major concern of project managers. Slippage occurs one day at a time and project managers need to be ever vigilant to keep slippage from accumulating to an unacceptable level. Since slippage is not a problem where there is sufficient schedule slack but is a problem where there is either insufficient or no slack, project managers cannot wait until slippage has become conspicuous. Rather, they need to have early warnings of potential schedule slips. These can be gleaned from the behavior of project staff by understanding the causes of slippage.

Slippage can be caused by complacency or lack of interest, lack of credibility, incorrect or missing information, lack of understanding, incompetence, and conditions beyond one's control, such as too much work to do. Project managers need to be on the alert to detect the existence of any of these factors in order to nip them in the bud before they result in schedule slips. (The existence of one or more of these factors does not guarantee that a schedule will be slipped. It merely provides the basis for a schedule slip and at the same time provides project managers with early signs of possible slips.)

B. Preventing Schedule Slippage

Schedule slips are not inevitable; in fact there is much project managers can do to limit them. They can, for example, subdivide tasks and assign responsibilities, establish check points and obtain timely and meaningful reports, act on difficulties, and finally show that it matters. Together, these steps will counter complacency, lack of interest, and lack of credibility. They will also provide opportunities to correct erroneous information and supply missing information, to build understanding, and to detect incompetence and conditions beyond the project staff member's control. If done seriously and with reasonable thoroughness, project managers stand an excellent chance of meeting their target schedules, assuming of course that they are realistic. There is nothing project managers can do to meet truly impossible schedules!

C. Impossible Schedules

Project managers should not agree to schedules they know are impossible to meet. If, however, they agree to schedules which they originally thought possible to meet but later discover are not possible, then they need to renegotiate their commitments or schedules as soon as they detect the difficulties. To do any less is to be unfair both to their customers and their project staffs.

IV. COST CONTROL

Cost control consists of assuring that work elements are accomplished within their respective budgets. Because of their differing characteristics, it is useful to have three separate budgets for each work element: a budget for direct labor, a budget for support services, and a budget for purchased services, materials, and equipment.

A. Budgets

1. Direct Labor Budgets

Expenditures for direct labor occur in relatively small units, perhaps as small as a fraction of an hour, and may be spread over a large number of people who are directly or indirectly responsible to the project manager. The expenditures are typically not coordinated at the level where they are incurred, so the project manager must establish individual budgets for the various work elements in order to be sure that their total does not exceed the total budget available for this work. Otherwise, minor overages for each individual element will accumulate to a major overage for the project manager, one that exceeds any reasonable contingency allowance that might exist.

2. Support Services Budgets

Expenditures for support services need to be handled separately because these functions are typically provided on a time-and-materials basis with no budgetary discipline. Support service groups will generally do whatever is asked of them and will charge accordingly, making no attempt to absorb costs due to errors of estimation or performance. Moreover, their work tends to be done in large lumps so that their overages are also lumpy and come without much warning. Thus, the opportunity to revise their efforts in order to accommodate their mistakes and deficiencies is limited. However, work assigned to support services tends to be more familiar than exotic, and it can be characterized and scoped quite well if the effort is made to do so. Accordingly, project managers should insist on obtaining detailed budgetary estimates for work to be done by support groups so that they can be scrutinized for completeness and accuracy. Since there is little chance to adjust budgetary allocations for support services later, they must be prepared with care at the beginning.

3. Budgets for Purchased Items

Expenditures for purchased services, materials, and equipment may be made on a basis ranging from time and materials to firm fixed price. The former need to be budgeted with the same care as support services, for the same reasons. The latter may require less detailed information for budgeting purposes, but they have other problems.

For example, suppliers may fail to deliver because they cannot perform for the price agreed to. Or suppliers may try to renegotiate the price for the same reason. Or unacceptable work or material may be supplied to meet the agreed price.

Suppliers might not provide timely notification of any difficulties, even when required to do so by contract. Also, contracts to suppliers are difficult to adjust unilaterally, and commitments to suppliers are thus harder to break than internal arrangements.

For all these reasons, budgets for services, materials, and equipment to be procured from suppliers should be identified separately and carefully prepared.

B. Expenditure Reports

Once budgets have been suitably prepared for the various components of the project, expenditures should be reported against them periodically. In the case of purchased services, materials, and equipment, the figures reported should be funds committed rather than invoices received. To wait for an invoice is to harbor a false impression about how much money is left to accomplish the remaining work.

C. Expenditure Audits

Expenditures should be audited to verify (1) that they refer to legitimate charges to the project and (2) that the work to which they refer was in fact done. Strange as it may seem, project budgets are occasionally charged for work that should have been charged to another project and for work which is yet to be done. The first type of false charge may be simply an error. Both types could be the result of chicanery on the part of an unscrupulous staff member. A project manager cannot afford to have the project budget pilfered by either type of false charge. Expenditure auditing will help curb these abuses.

D. Comparing Expenditures Against Budgets

After expenditures have been checked to verify that they are legitimate, they should be compared to budgets and to the work accomplished. The comparisons can take any of several forms. One possibility is to compare expenditures for each period to the budget for the period. This approach indicates how well the project is going currently, but it says little about how well it is going overall. To overcome this shortcoming, cumulative expenditures to date can be compared to the cumulative budget to date. This latter approach is particularly useful to show if the project is in more or less budgetary difficulty than it was in the preceding period. It is also useful in determining the extent to which an accumulated overage is within the budgetary contingency allowance.

Graphs are often helpful when comparing cumulative expenditures against the cumulative budget. Figure 4.3 shows a typical cumulative budget curve on a linear scale, and Figure 4.4 shows the same information on a semilogarithmic scale. Each scale has its advantages as a control device.

The linear scale provides the same precision at the end of the project as the beginning, that is, very good precision, and it is at the end that the project manager worries about small amounts and wants the precision. The semilogarithmic scale lacks this precision at the end of the project.

On the other hand, the semilogarithmic scale easily accommodates parallel percentage warning lines and danger lines. Since a semilogarithmic scale is a ratio scale, a line drawn parallel to it is a fixed percentage different from it. The project manager can thus easily draw warning lines, say, ±5% of the budget line, and danger lines, say, ±10% of the budget line. The actual percentages used depend, of course, on the amount of the budgetary contingency allowance. The warning percentage should be small compared to the total contingency allowance. The danger percentage should be, say, a third or half of the total contingency allowance.

A project manager (or a clerk) can plot the cumulative expenditures on the same semilogarithmic graph as the budget, warning, and danger lines. When cumulative expenditures fall off the budget line but within the warning lines, the project manager need not be too concerned. The contingency allowance is sufficient to cover such discrepancies. However, when cumulative expenditures fall outside the warning lines, the project manager should find out what is going on. The discrepancy between budgeted and actual expenditures is getting large enough to suggest that something is awry. Moreover, it is growing in absolute terms if

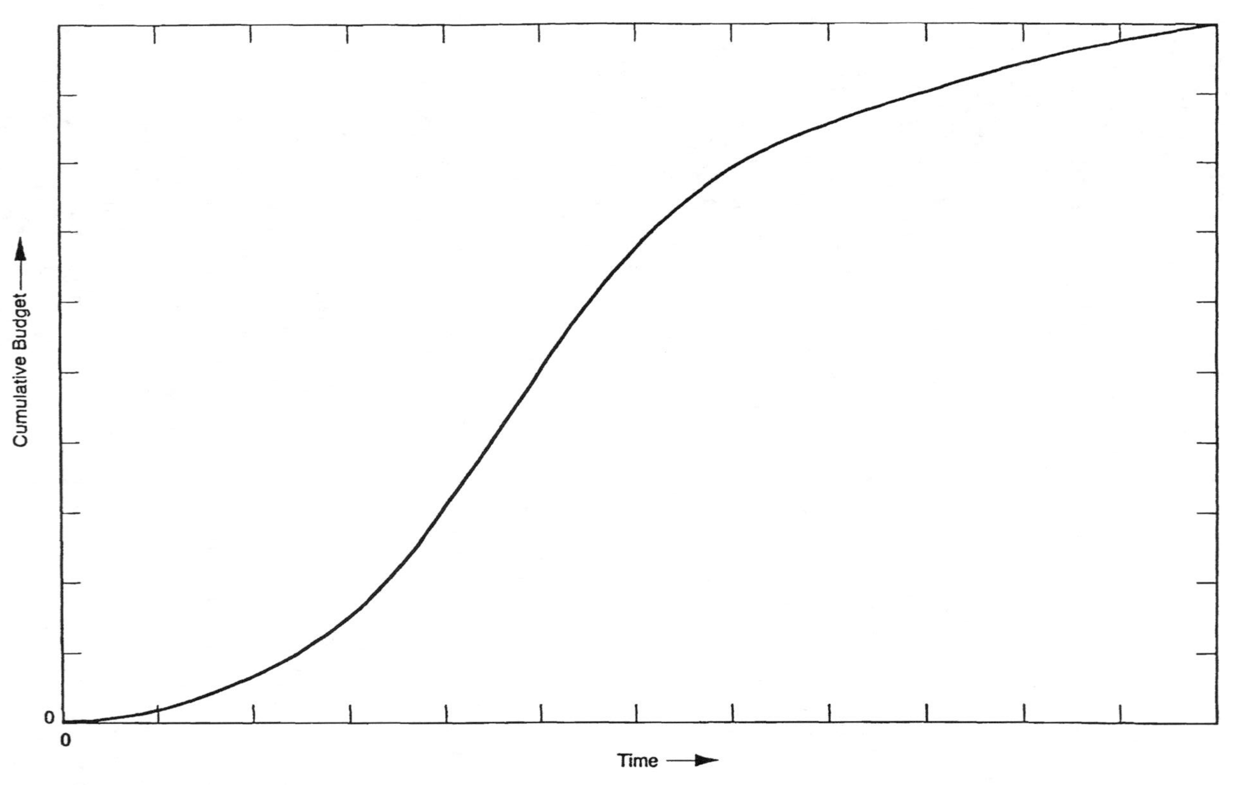

Figure 4.3. Budget versus time for a typical project, linear scale.

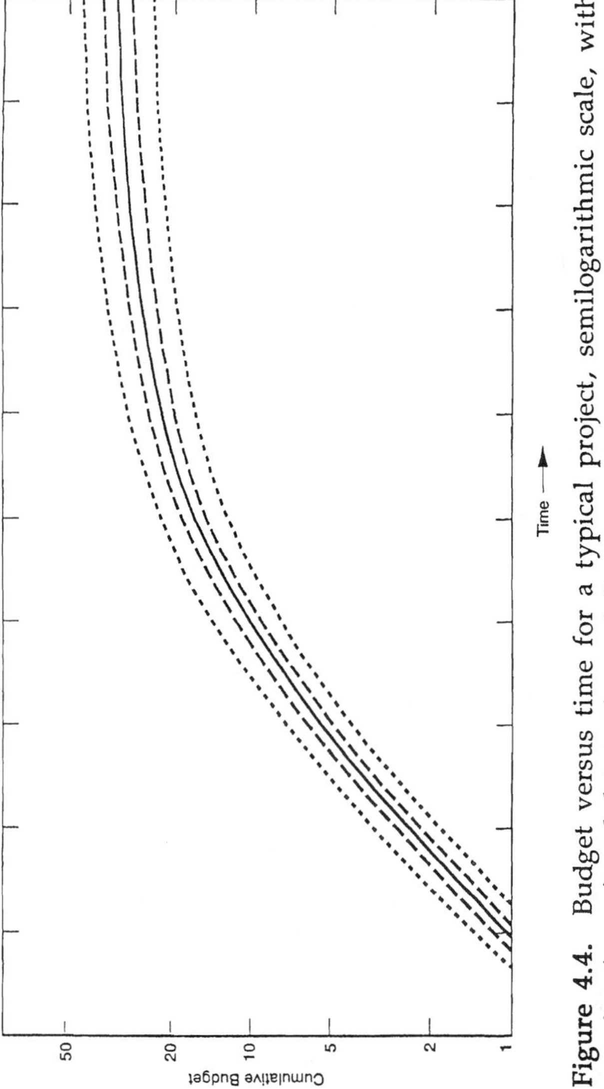

Figure 4.4. Budget versus time for a typical project, semilogarithmic scale, with warning (– – –) and danger (·······) lines.

it just stays at the same percentage above or below the budget line.

E. Overspending

When cumulative expenditures exceed the upper danger line, the project manager may have a serious problem to resolve: A significant portion of the contingency allowance may be depleted before the project is finished. This requires the project manager to reprogram the remainder of the project in order to conclude it without overrunning the total budget.

F. Underspending

So far, one might have assumed that an excess of cumulative expenditures over budget is more serious than a deficiency. However, a deficiency, especially in the early or middle phases of a project, can indicate significant trouble: if the budget is not being spent, then some work is probably not being done. Not only does this endanger completion of the project on time, but it also threatens the budget because a compressed work schedule almost always entails inefficiencies and increased costs.

G. Budget Needed to Complete the Work

Another expenditure/budget comparison is both the most useful and the most difficult to make—a comparison of the budget yet unspent with the work yet to be done. This comparison is truly the most useful because it indicates whether the work remaining can be done within the available resources. The other comparisons are in fact retrospective and require the project manager to infer what future conditions will be by examining the past. A comparison of the work to be done and the budget yet unspent is more direct in terms of helping the project manager gauge and influence the future. Thus, this comparison is more helpful than the others. At the same time, it is difficult to use because it repeatedly requires re-estimates of the budget needed to complete the project.

Each of the expenditure/budget comparisons has its advantages and disadvantages, and wise project managers will use each of them appropriately. Cumulative expenditures will likely be the easiest to apply overall and will probably serve as the basic approach. Periodic expenditures are useful for tracking selected aspects that have a tendency to go awry. And comparisons of work to be done versus budget yet unspent will help project managers restructure their plans if necessary.

V. EARLY WARNING SIGNALS

Early warning signals indicate when a project or project segment is headed for trouble even before a control issue arises. They enable project managers to prevent or at least limit the damage that might otherwise occur, thus minimizing the amount of correction needed later. Two types of early warning signals are

1. Periodic assessments (e.g., weekly assessments) based upon easily obtained quantitative project metrics
2. Reviews, i.e., third-party inquiries into the qualitative nature of various aspects of the project

A. Periodic Assessments

Three straightforward project metrics are

1. The ratio of (a) the percent of the schedule and budgetary contingency allowances remaining to (b) the percent of project or project segment remaining *versus* time
2. The ratio of (a) the cumulative number of substantially equal-sized work packages scheduled for completion by

a given date to (b) the number actually completed correctly by that date *versus* time

3. The number of change orders, anomalies, discrepancies, and other ad hoc work items open divided by their respective rates of closure *versus* the amount of time remaining for their completion

Each of these metrics can be gathered with relative ease and provides early warning of later problems.

For example, if the ratio of percent of a contingency allowance consumed to percent of work accomplished is greater than 1.0 and growing with time, the project will likely run out of the allowance before the work is done. This means that no allowance will exist to correct mistakes that occur late in the project, which is unsatisfactory. Disproportionately rapid consumption of a contingency allowance signifies that the project manager should replan the project, which may necessitate renegotiating the scope, schedule, or budget with the customer.

Similarly, if the ratio of cumulative number of work packages scheduled for completion by a given date to the number actually completed correctly is greater than 1.0 and growing with time, the schedule will likely expire before all the work is done, again an unsatisfactory outcome. As before, the project manager should replan the remainder of the project in order to have a compatible set of scope, schedule, and cost objectives.

Likewise, if the closure rate of open ad hoc work items is insufficient to close them in the time remaining for their completion, the project will likely not meet its objectives. Here, too, the project manager should replan the project in order to have and meet a compatible set of objectives.

None of these data is especially difficult to gather, but they are not acquired for free. Some small effort is needed to do the clerical work. This effort should be more than repaid by being able to fix problems while they are still small rather than waiting until they are larger and less tractable.

A more significant cost is the managerial effort required to scrutinize the data, investigate causes, and plan remedial action. Yet this effort must be expended for two reasons. First, if no effort is made to determine and fix problems when they are small and manageable, they will likely increase in difficulty. Second, if the data are collected but not used, the project manager will appear uninterested in preventing problems. Others may interpret this apparent lack of interest to mean that they need not be diligent, resulting in even bigger problems than would otherwise occur.

The metrics discussed apply to any size of project element, from the overall project to the lowest project level at which a contingency allowance or system margin is held, at which work or task completions are defined, and at which ad hoc work items occur. This fact means that allocating allowances and margins, defining work products (or tasks), and identifying ad hoc work items at low levels of the project hierarchy are worthwhile not only in terms of accountability, but also in terms of project monitoring.

Admittedly, allocating allowances and margins to low levels makes it difficult to subsequently shift an allowance or margin from an area where it is not needed to another area where it is needed. However, this difficulty is not insurmountable, as described in Chapter 5 on risk management, and it should not be allowed to interfere with monitoring a project at various levels in order to detect and correct problems early. The metrics described should be applied at the lowest levels of the WBS possible in order to detect and correct problems as soon as possible. Early detection and correction are key to low-cost solutions, which will also minimize the likelihood of problems overall.

B. Reviews

Reviews are typically held at critical junctures in the course of a project, for example, upon conclusion of requirements definition, upon conclusion of preliminary design, and so forth. Moving forward prematurely at one of these junctures could mean a bad outcome, and the purpose of a review is to obtain an early indication or prognosis of the outcome if work were to proceed. Two types of inquiry are used to identify potential problems. One assesses the work done: is it sound enough to enable the project to build upon it? The second assesses arrangements for proceeding: are the project and its supporting elements ready and are the risks of proceeding acceptable? Each type is now discussed in turn.

1. Assessment of Work Done

1. *Was the right work done?* In deciding whether the right work was done, reviewers should consider assumptions that underlie the work, the results of trade-off studies, and decisions that have been made. They should look for evidence that the assumptions are realistic and reasonable, that the range of options has been appropriately explored, and that due attention was paid to the results of the trade-off studies when decisions were made.
2. *Was the work done right?* In deciding if the work was done right, reviewers should examine available quality assurance information and judge whether the work conforms to the plan or design and whether the resulting products are likely to perform as intended.
3. *Will putting products under configuration control help or hurt?* In deciding whether configuration control will help or hurt, reviewers are effectively deciding if the work done to date is mature and stable enough to warrant others to rely on it. Configuration control imposes formal procedures (and concomitant delays) for notifying others of proposed changes and for evaluating their pros and cons before authorizing them. Its purpose is to protect those depending on an item from surprise and perhaps undesirable changes. If an item is known to be subject to many changes, placing it under configuration control may be premature. On the other hand, the ease

of changing an item not under configuration control must be weighed against the disadvantage of others not *necessarily* being consulted about proposed changes and therefore not being able to rely upon a well-defined baseline.

Detailed questions to help assess work done on system development projects are given in Appendix E.

2. Assessment of Readiness to Proceed

1. *Is the overall plan for future work suitable?* In assessing the suitability of an overall plan, reviewers should decide if it is based upon realistic assumptions and upon consideration of technical and organizational options, such as make versus buy; one phase or multiphase (incremental development or delivery); and parallel versus serial work to deal with technical, schedule, and resource risk.
2. *Are the detailed plans for next phase appropriate?* In assessing plans for the next phase, reviewers should decide
   a. If the objectives of the next phase are clearly stated, correctly understood, and translated into specific deliverables
   b. If the critical technical and organizational issues are identified and suitably provided for
   c. If the people who will do the work participated in preparing the budgets and schedules, believe that they are realistic, and are committed to them
   d. If the mechanisms that the manager will use to develop and manage a responsive and responsible team will be effective.
   e. If the communication mechanisms and interactions among team members and with others are described sufficiently and will be effective
   f. If the metrics proposed for gauging progress and obtaining early warning of difficulties will be effective

g. If line management and project management have concurred in any make-or-buy decisions
h. If the project and the customer have concurred in any decisions regarding phased deliveries
i. If line managers have committed, but not overcommitted, their personnel to the project

Finally, reviewers should examine the periodic assessment data described above and ask about any problems that the project manager may be able to explain. Their aim here is to determine any systemic conditions that need to be remedied but have not yet been mentioned, due either to myopia or to an unwillingness to face unpleasant facts.

5

Risk Management

Projects contain uncertainties and thus involve (1) the probability that something could go wrong and (2) adverse consequences that will occur if it does go wrong. The product of these two factors is called *risk*, and taking precautionary steps to minimize risk is called *risk management*. More specifically, *risk management* refers to assessing, reducing, and controlling risks so that project objectives are met without requiring excessive schedules or resource budgets.*

In a sense, all of project planning and control can be considered risk management. In this chapter, however, we focus on techniques that extend beyond ordinary planning and control. These techniques become increasingly important as project customers demand ever better project results.

For projects that are very similar to others, basic methods of estimating, planning, and control have likely already

---

*The insurance industry uses the term risk management to mean pooling or sharing risks so that one entity does not bear the full consequences of specified adverse events. We are not addressing this use of the term here. Later, however, we refer to transferring risk from one person or entity to another, which is the essence of insurance.

been adjusted to accommodate risks automatically. In these cases, the original risk analyses and the development of appropriate risk management techniques may now be lost, but the methods work well enough anyway. Under these conditions, conscious thinking about risk management may not be required.

For many projects though—especially development or leading-edge projects—significant analysis may be required to understand and manage project risks. This chapter, therefore, describes a way to assess, reduce, and control risks when a conscious, semiformal process is appropriate.

This discussion is limited to the risks of accomplishing the project. It does not extend to any liabilities associated with poor design such as software that fails or a structure that collapses once the project has been completed. However, project team members who receive products from others should always take the reliability of these products into account when assessing their own risks.

I. RISK ASSESSMENT

Risk assessment consists of identifying risk items and estimating their consequences.

A. Risk Item Identification

A risk item is any technical performance, resource, or schedule outcome whose likelihood of being satisfactory is judged a priori to be less than 100%. This judgment is made after taking account of tests and other factors that may already be planned to find and fix errors of omission and commission.

Overall project risk is based upon two types of risk at the next lower level of the project's work breakdown structure (WBS):

1. Risks concerned with interfaces between elements at the next lower WBS level
2. Risks concerned with the elements themselves at the next lower WBS level

The risk items in the second of these cases are also of the same two types, and so forth. This recursive relationship continues to the lowest level of the work breakdown structure. Thus, assessing project risk requires judgments about the risk of every WBS element.

B. Risk Consequence Estimates

Adverse consequences for the risk items identified may be in terms of performance shortfalls, budget overruns, or schedule overruns. With the help of project team members and other experts, project managers should first make an outside estimate of the adverse consequences for each risk item and then put these estimates on a common denominator for comparison. It is convenient to put all consequences on a monetary basis, as follows:

1. The monetary equivalent of a performance shortfall is the additional cost that is necessary to achieve the required performance (assuming that sufficient time is available).
2. The monetary equivalent of a schedule overrun is the additional cost that is necessary to meet the schedule (assuming that performance is not an issue).
3. The monetary equivalent of both a performance shortfall and a schedule overrun is the additional cost that is necessary to both achieve the required performance and meet the schedule.

II. RISK REDUCTION

Risk reduction consists of reducing uncertainties, reducing consequences, avoiding risks, and transferring risks.

A. Uncertainty Reduction

Reducing a risk's uncertainty may have either of two effects. It may convert the risk to a certainty, to be dealt with accordingly. Or, it may merely reduce the probability of an adverse outcome.

Uncertainty may be reduced by the following methods:

1. *Prototyping, simulating, and modeling.* These three methods share the notion of using a representation to investigate selected aspects of requirements, a design, or a plan in order to be more certain about their suitability.

   A *prototype* is a mock-up or representation of (only) the areas of investigation in order to test its acceptance, for example, a colored, scaled drawing of a proposed computer display. A *simulation* is an imitation of the functioning or behavior of one system by means of the functioning or behavior of another, for example, a computer or algorithmic simulation of a complex industrial process. A *model* is a miniature representation of physical relationships, for example, a scale model of complex piping to test spatial relationships.
2. *Planning in detail.* Detailed planning means detailed estimating, budgeting, and scheduling, including resource scheduling, so that it will be clearer in advance whether or not the project can be accomplished as contemplated overall.
3. *Parallel alternative developments.* A parallel alternative development is essentially a concurrent backup. However, unlike a backup that is pursued only after an adverse outcome has occurred and whose cost might be avoided, a parallel development represents costs that will necessarily be incurred. It can be worthwhile, though, especially when time is more important than cost.

   This reduction in uncertainty is really a reduction in the likelihood of failure. The combined or joint probability of all alternatives failing is the product of the individual alternatives' likelihood of failure. Thus, since each alternative has a likelihood of failing that is less than 1.0, their product is a number that is smaller than any one of the individual likelihoods of failing (assuming that their failures are mutually independent).

4. *Checking references.* Checking references can reduce uncertainties about individual abilities or inclinations (but does not by itself improve them).
5. *Using trained or certified staff.* Training and certification can establish minimum competencies for staff and reduce downside uncertainties about them.
6. *Using proven technology.* Technologies that are proven in comparable situations have lower a priori uncertainties about their suitability than unproven technologies (although the latter could turn out a posteriori to be more suitable).
7. *Verifying the suitability of inputs.* Tests and inspections can be used to verify that inputs, whether information or materials, are correct and complete so that subsequent work will not be wasted.

B. Consequence Reduction

A risk's consequences can be reduced by the following methods:

1. *Decoupling related items.* Decoupling refers to removing dependencies. For example, if one person is to work on two tasks in series, then lateness in completing the first task can cause lateness in completing the second task. If the assignments are decoupled and each task is assigned to a different person, then lateness in completing the first task does not ipso facto cause lateness in completing the second.

   Decoupling can often be used to reduce risk consequences when resources are scheduled, when information or technology are shared, and when two items are derived from a common process.

2. *Providing margins or reserves.* Margins and reserves are resources held for the express purpose of accommodating mistakes ("oops") and oversights ("I forgot"). Sometimes they are called contingency allowances. Margins and reserves are held to fix problems that arise internally, not to handle requests from outside to do more, better, or differently. Such externally generated needs should be accompanied by sufficient additional resources to accommodate the requested changes.

   What is internal to one level can be external to the next lower level. For example, Level IV fixes its own problems with its own margins and reserves and responds to Level III's requests for changes by asking Level III to provide additional resources if needed. Level III provides these resources from its own margin or reserve.

   To forestall lower-level managers from holding excessive amounts of margin or reserve, managers at higher levels may hold all margins and reserves on behalf of their constituent lower-level items. In this case, each lower-level manager must have easy access to a margin or reserve when needed. However, a higher-level manager may decide to accept a particular mistake or oversight in order to allocate the margin or reserve elsewhere, where need is even greater.

   On the other hand, each person responsible for a WBS item is more likely to feel fully accountable if margins and reserves are distributed to them individually to steward. This approach, however, requires a decision about how much of the overall margin or reserve should be allocated to each person. A reasonable scheme is for the project manager to retain a third or half of the total and to pass the remaining portion to the next level of WBS element managers. Allocation among these managers can be proportional to their respective risk reduction costs. These managers can then retain a third or half of their allocations and pass the remainder to their own next-level managers, again in proportion to their risk reduction costs.

When margins or reserves are partially allocated to lower-level managers, some of the lower-level managers may find in time that their allocations are too small while others find that their allocations are more than sufficient. In this event, the project manager needs to sweep up or retrieve part of the unused allocations, say half of them, and redistribute the available margins and reserves to the more needy elements. Those who relinquish margin or reserve in this way may then later have problems that require some margin or reserve to be restored. The project manager should readily grant this restoration.

Retrieval and restoration are based on the premise that not everyone will need his or her part being retrieved, but some will. By retrieving and aggregating a portion of the unused allocations, margins and reserves can be made available where needed, including to WBS elements that appeared before retrieval to have an excess.

C. Risk Avoidance

Some risks can be avoided by reducing requirements or defining them more completely, increasing the budget, or extending the schedule. Budget increases and schedule extensions may not be available overall, but they can sometimes become available locally by changing scope elsewhere in the project or by accomplishing preceding or succeeding tasks in less time. Adjusting preceding and succeeding tasks may require more resources, more intensive use of resources, reductions in requirements, or a combination.

D. Risk Transfer

A risk can sometimes be reduced for the element at hand by transferring it to another entity that can better bear it. Two methods of transferring risk are the use of insurance and the alignment of responsibility and authority.

1. *Insurance.* Insurance, even so-called self-insurance, transfers risk to a larger entity that can bear the risk. Each element contributes to the insurance pool in some fashion. Contingency allowances or reserves are one version of insurance *when they are pooled.* Paying a premium to a third party, that is, an insurance company, is another version. Not all problems can be solved with insurance, especially when time has run out.
2. *Aligning responsibility and authority.* Another way to transfer risk is to align responsibility and authority so that decisions are made by those who have to carry them out. Such alignments generally result in reduced adverse probabilities and consequences, for four reasons:
   a. Decision makers tend to judge external circumstances realistically when they know that they will have to carry out the decisions personally.
   b. Decision makers can judge their doers' capabilities realistically when the two are the same person.
   c. Misunderstandings between decision makers and doers are minimized when the two are the same person.
   d. People who make their own commitments are more likely to "go the extra mile" to fulfill them than those who do not make their own commitments.

With the help of project team members and other experts, the project manager can evaluate each risk item in terms of its potential for and cost of uncertainty reduction, consequence reduction, avoidance, and transfer. The best method of risk reduction can then be selected for each item in terms of risk reduced per unit cost.

III. RISK PLANS

A risk plan shows what risks will be reduced, and how, for given total levels of expenditure. The plan starts with risks whose consequences are absolutely unacceptable and must be reduced at least to acceptable levels. The plan then addresses risk that will be reduced by applying resource *a priori*. The plan finally addresses remaining or residual risk items that will be accommodated by providing margins and reserves (contingency allowances).

Some project managers rely exclusively on margins and reserves to handle all risks. They do not assess and reduce risks before the fact and thereby miss important opportunities to economize on the use of project resources. Worse, by not knowing confidently where or when resources will be needed, these managers complicate resource scheduling both for their own projects and throughout the organization. Accordingly, we do not recommend relying exclusively upon margins and reserves to handle risks.

IV. RISK CONTROL

Risk control consists of monitoring the use of margins and reserves, reassessing risks, replanning risk reduction as needed, and evaluating the impact of proposed changes on risk.

A. Margin and Reserve Monitoring

The manager of each project element that has a margin or reserve (contingency allowance), should periodically calculate the ratio of (a) the percentage of margin or reserve consumed to (b) the percentage of work successfully accomplished. The percentage of work successfully accomplished is based on current (re-)estimates of total work required, that is, work done completely and correctly plus current (re-)estimates of work to be done.

B. Risk Reassessment

When a project element's ratio of percentage of margin or reserve consumed to percentage of work successfully accomplished exceeds 1.0 by more than, say, 5% the manager should analyze the causes and reassess the risks. The latter includes identifying risks not previously identified.

C. Risk Reduction Replanning

Based upon the results of risk reassessment, managers should revise their risk reduction plans as needed.

D. Risks Versus Changes

Project element managers should assess risks introduced (or re-introduced) by proposed changes and revise their risk reduction plans as needed. Proposed changes should always be examined carefully because they often are the basic cause for matters getting out of hand and raising the frequency and severity of adverse outcomes.

Project risk management is not performed in a vacuum. Rather, it is performed in each case in a specific context and requires keen insight into the content of each project element. But knowledge about the context and the content of a project element is not enough. Each project element manager needs also to know the mechanics of assessing, reducing, and controlling risks described in this chapter.

6

Coordinating and Directing Techniques, Part I

Effective coordination and direction of the project team depend essentially on successful motivation and good communication and interaction. Successful project managers understand what makes their team members tick and how to communicate with them productively. Accordingly, our approach in this chapter and the next is to present principles of motivation and effective communication and interaction and show how they can be applied in project activities. This chapter focuses on both general aspects and individual aspects of motivation, while the next focuses on communication and interactions.

I. GENERAL ASPECTS OF MOTIVATION

The most potent force that a project manager has to elicit cooperation from team members is to genuinely care for them. This is not manipulation. Rather, it is behavior that serves mutual best interests. When the project manager takes the initiative in caring for team members, he or she can expect them to reciprocate by caring about the project and its manager and cooperating within their ability to do so. This is also true in the project manager's relationships with superiors, peers, support staff, and customers.

Caring can be exhibited in many ways. Common everyday courtesies indicate caring and should be extended no matter what the other's station. Caring is also shown in conversation when the listener pays attention to the speaker and tries seriously to understand the speaker's point of view and feelings. When in a group, caring is shown by those who solicit the views of others who have not been able to insert themselves into the discussion.

Another aspect of human interaction is the practice of paying compliments. Compliments are positive comments, or *strokes,* and enhance the other person. As a result, people who are complimented feel good about themselves and about those who paid the compliment. They are thus inclined to cooperate with those people if possible. Of course, compliments must be sincere. If they are not, their insincerity will be detected and cooperation may in fact be withheld.

There are two kinds of positive strokes: conditional strokes and unconditional strokes. A conditional stroke is a compliment that can be paid because of something the recipient did. It is called a *doing stroke.* An example is "You did a nice job on that report."

An unconditional stroke, in contrast, is a compliment that can be paid just for being, and no strings are attached. It is called a *being stroke.* Examples are "It's good to see you" and "I'm glad to have you on the project."

Of the two types of positive strokes, unconditional strokes are the more potent. That is, they affect one's sense of self-worth more deeply and pervasively.

While discussing stroking, it is appropriate to caution against certain counterproductive comments. Negative strokes are putdowns and diminish the other person. An example might be "We'd like to pay you what you're worth, but the minimum wage law won't let us." Negative strokes are resented and build animosity. They should never be used. Negative strokes are not to be confused with criticism, which is discussed in Chapter 7, Section II.

Closely related to negative strokes are blurred strokes. A blurred stroke leaves the recipient wondering what was meant. Was the stroke a compliment or a putdown? An example is "Your work isn't nearly as bad as they said it would be." There is always a negative quality to a blurred stroke, and it should not be used.

Similar to blurred strokes are crooked strokes, which also have a negative component. They appear on the surface to be compliments but they are actually very sarcastic putdowns. An example is "You look nice today; your socks match." Crooked strokes should be avoided. Even if one receives a crooked stroke, one should not return it.

Finally, the project manager should be aware of the consequences of ignoring an individual, that is, not giving that person any strokes. People who receive no strokes are likely to foul things up just when it hurts most. They do this to get attention. Their behavior is similar to that of a child who does not normally receive some attention. The child then manages to get attention by misbehaving.

Because unfortunate consequences can result from nonstroking, project managers cannot afford to focus only on poor performers. If satisfactory and outstanding performers are ignored, one or another of them will foul up just to receive a share of the manager's attention.

If stroking patterns among a project manager's team members are unhealthy, they can and should be changed. The best way for a manager to do this is to personally practice caring behavior, including positive stroking. Not only will this caring behavior enhance the team members' feelings about themselves, but it will also set an example for them to emulate in dealing with each other. In addition, the project manager can help improve stroking patterns by accepting a positive stroke graciously whenever one is paid and by not stroking another's bad behavior.

It was mentioned above that compliments must be sincere and paid genuinely if they are to be accepted as evidence of caring. Such sincerity grows out of an overall fabric of straightforward communication. One should not expect a compliment to be taken seriously if everything else that one says is fuzzy or misleading.

Straightforward communication, sometimes known as straight talk, occurs when the speaker (or writer) says what he or she means. For example, straight talkers say *I* when they mean *I* and do not hide behind vague *we*s or *they*s. They are specific when they can be. If they need something by Tuesday noon, they do not say "anytime next week" (and then perhaps become upset if the item is not ready by Tuesday noon). If they mean no, they say "no" at the time. They do not just merely discourage the other person, hoping that he or she will get the point without being told no. And straight talkers ask for what they want, rather than hint about it while hoping that the listener will deduce what they want.

II. INDIVIDUAL MOTIVATION

Project managers must understand not only general aspects of motivation but also individual motivational differences if their project teams are to realize their full productivity potential. This section discusses three models of human behavior that can help project managers capitalize on individual motivational strengths and compensate for shortcomings: the Porter strength deployment model, the Myers-Briggs Type Indicator®, and the Casse communication styles model.

A. The Porter Strength Deployment Model

Elias Porter developed a model of how different individuals behave when everything is going well and how they behave when they face opposition and conflict. If everything automatically went well in projects, we would not need to be concerned (nor need this book), for there would be no problems. Projects are, however, characteristically beset with problems, so it pays to understand how different people react in these circumstances and how to make the best of the situation.

At the core of Porter's model are two basic assumptions:

1. People act in identifiable ways toward others because they expect that doing so will prove gratifying to themselves. They temper or adjust what they do according to what they expect the other person to do, but in all cases they act in ways that they believe will result in a personally rewarding interaction.
2. The motivations or sources of gratification of others can be inferred by observing how they behave. That is, their observable behavior is consistent with their motivations. We all commonly accept this assumption because we all infer how someone else will act in a particular circumstance based on our prior experience with that person. We are not always correct, but our errors are more likely to result from not fully knowing everything about the person or the circumstances than from the assumption being wrong.

With these two assumptions, Porter identified three archetype motivations for human interactions. Each individual's behavior represents a blend of these motivations. Individuals differ in their blends just as they differ in their fingerprints. The archetype motivations and significant motivational blends are described in Table 6.1 and shown in Figure 6.1.

Table 6.1. Archetype Motivations and Motivational Blends

Archetype motivations:
- Altruistic-nurturing motivation
  - Enhancement of the other person's welfare is paramount.
  - Characteristic strengths: see Table 6.2.
- Assertive-directing motivation
  - Achievement of goals by influencing others is paramount.
  - Characteristic strengths: see Table 6.2.
- Analytic-autonomizing motivation
  - Achievement of self-reliance, self-suffiency, self-dependence is paramount.
  - Characteristic strengths: see Table 6.2.

Motivation blends:
- Assertive-nurturing motivation
  - Assertive in helping others; not autonomy-seeking.
- Judicious-competing motivation
  - Strategic; the power behind the throne; not nurturing.
- Cautious-supporting motivation
  - Helpful and self-sufficient; not assertive.
- Flexible-cohering motivation
  - Seeks inclusion in group activities; gratified by doing whatever the situation calls for.

Source: Adapted from *Leader's Guide to Relationship Awareness Training*, Personal Strengths Publishing, Inc., P.O. Box 397, Pacific Palisades, CA. Used by permission of the publisher.

The archetype motivations (called *altruistic-nurturing. assertive-directing,* and *analytic-autonomizing*) are at the corners of the triangle in Figure 6.1. Blends of approximately equal parts of two archetype motivations (called *assertive-nurturing, judicious-competing,* and *cautious-supporting*) are represented by areas along the sides of the triangle. And a blend of approximately equal parts of all three archetype motivations (called *flexible-cohering*) is represented by the central area of the triangle.

The closer an individual's blend is to a given corner of the triangle, the more dominant the corresponding archetype motivation is overall. This relationship is shown by the

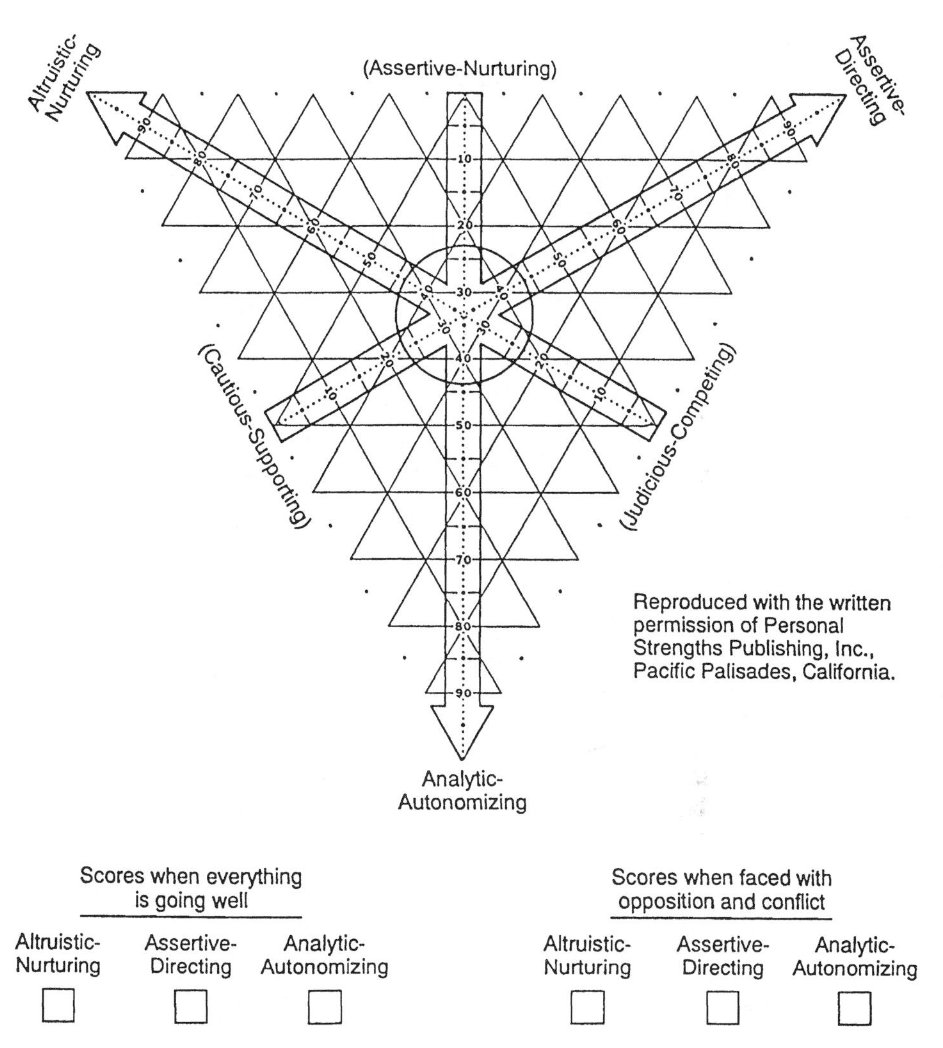

Figure 6.1. The Porter strength deployment interaction triangle.

Table 6.2. Characteristics Strengths and Weaknesses of Porter's Archetype Motivations[a]

Strength	Weakness
Altruistic-nurturing motivation	
Trusting	Gullible
Optimistic	Impractical
Loyal	Slavish
Idealistic	Wishful
Helpful	Self-denying
Modest	Self-effacing
Devoted	Self-sacrificing
Caring	Smothering
Supportive	Submissive
Accepting	Passive
Assertive-directing motivation	
Self-confident	Arrogant
Enterprising	Opportunistic
Ambitious	Ruthless
Organizer	Controller
Persuasive	Pressuring
Forceful	Dictorial
Quick to act	Rash
Imaginative	Dreamer
Competitive	Combative
Risk-taker	Gambler
Analytic-autonomizing motivation	
Cautious	Suspicious
Practical	Unimaginative
Economical	Stingy
Reserved	Cold
Methodical	Rigid
Analytic	Nit-picking
Principled	Unbending
Orderly	Compulsive
Fair	Unfeeling
Perservering	Stubborn

(*continued*)

Table 6.2. Continued

Strength	Weakness
Flexible-cohering motivation	
Flexible	Inconsistent
Open to change	Wishy-washy
Socializer	Cannot be alone
Experimental	Aimless
Curious	Nosy
Adaptable	Spineless
Tolerant	Uncaring
Open to compromise	Lacking principles
Option-seeking	Lacking focus
Socially sensitive	Deferential

[a]The left entry in each pair is a characteristic strength. The right entry is the corresponding weakness, i.e., a strength overdone for the situation at hand (in the eyes of the beholder).

Source: Adapted from *Leader's Guide to Relationship Awareness Training*, Personal Strengths Publishing, Inc., P.O. Box 397, Pacific Palisades, CA.

numbers leading from the corner to the opposite side of the triangle. Thus, 100% coincides with the corner and 0% lies on the opposite side. It is no accident that the percentages of the three primary motivations total to 100 for every point in the triangle.

Porter developed a questionnaire called the Strength Deployment Inventory for individuals to use in determining their own motivational blends when everything is going well and when facing conflict and opposition. Most people have different motivational blends in the two types of situations, although some individuals are remarkably consistent. The shift from one blend to another can be represented by an arrow that points from the "going well" blend to the "conflict and opposition" blend, as in Figure 6.2.

People use their "going well" blends and "conflict and opposition" blends differently. When everything is going well, they support their dominant behavior or motivation, if they have one, with their auxiliary behaviors. That is, they have access to and can use all their behaviors at once. Simultaneous access to all behaviors facilitates doing whatever is appropriate to advance the group's efforts toward common objectives.

When facing opposition and conflict, on the other hand, people use their behaviors and motivations one at a time. Thus, they try to succeed using the behavior with the highest score in the "conflict and opposition" case. If this works, the conflict and opposition stop and the situation returns to "going well." If it does not work, then sooner or later this approach is abandoned and they try their next preferred behavior. Maybe it works, in which case the situation returns to "going well," and maybe it does not work. If it does not, then sooner or later it also will be abandoned and the least preferred behavior will be taken up, in the hope that it will save the day. Resorting to a least-preferred behavior may be done reluctantly and possibly with vengeance (for being forced into such a low-preference style).

Concerning this last point, people who do not use a behavior often are unlikely to get very good at it. When driven to using a least-preferred behavior, an individual may lack the finesse needed to use the behavior gracefully, due to a lack of practice. Often such an attempt is overdone and ineffective.

The astute project manager will try to infer how project team members are motivated and then use this insight productively. For example, the project manager can check whether the team has all the primary motivations represented to some degree. Is there at least one person who will just naturally analyze difficult situations, and one who is altruistic-nurturing and will naturally try to harmonize different viewpoints when necessary, and one who is sufficiently assertive and directing so that the group will not flounder for lack of leadership? If a needed primary motivation is

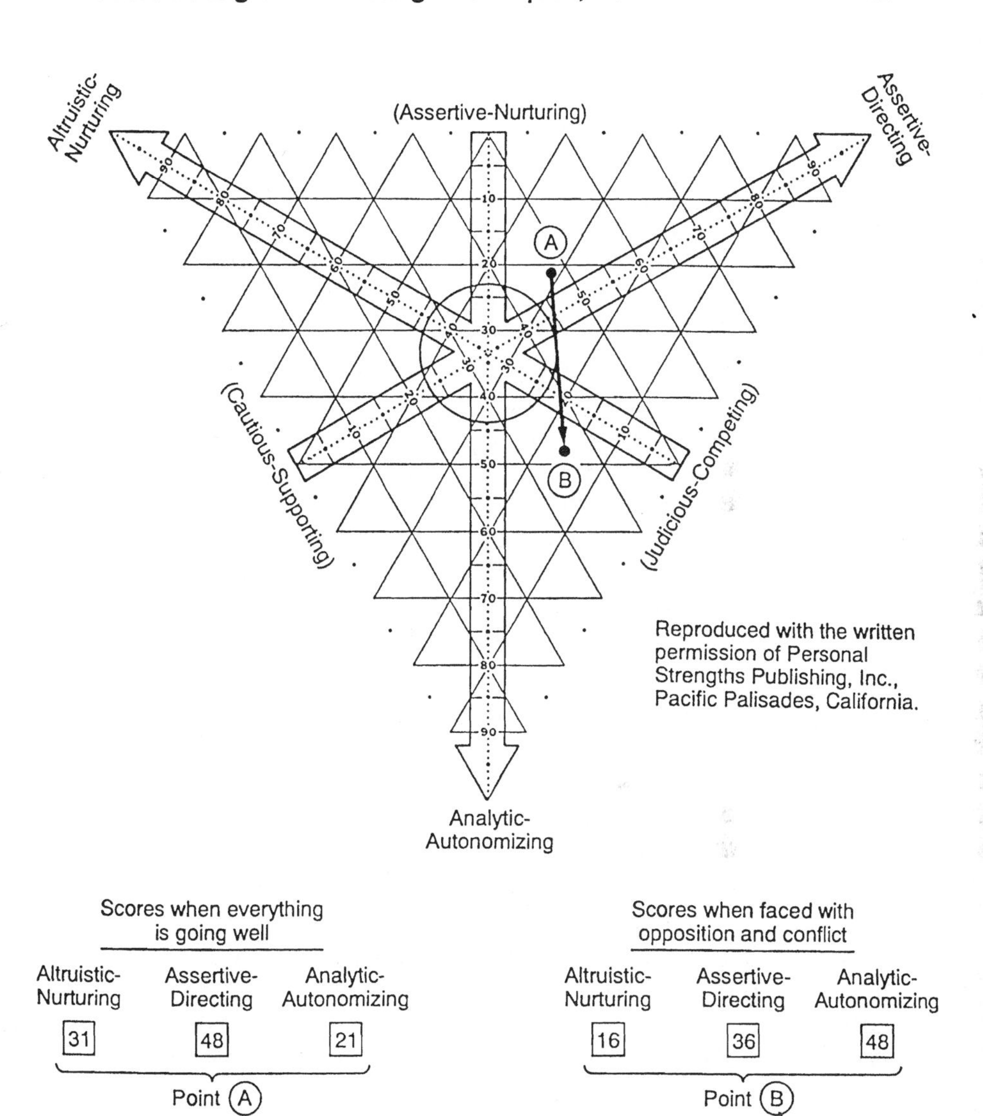

Figure 6.2. The Porter strength deployment interaction triangle for a hypothetical person.

missing, then the project manager can either consciously try to fill the void(s) personally, structure meeting agendas so that the needed activities occur even if they are not quite natural for anyone there, or bring in resource people who will supply needed ingredients.

Another possibility is for the project manager to use insight in selecting team members in the first place. Whenever project managers have a choice, they can try to obtain all the needed behaviors, as well as the technical expertise, by selecting individuals according to their motivations.

A third use of motivational insight occurs when disagreements exist about what to do. A wise project manager will keep team members from having to act according to their least preferred motivations and thereby avert vengeful, clumsy, and unproductive behaviors. Moreover, the project manager will try to avoid wasting everyone's time and effort by avoiding arguments that are basically rooted in individual behaviors and approaches rather than in differences in objectives. Thus, the project manager will accept individual ways of doing things as long as they do not cause a problem elsewhere and will also try to develop such forbearance throughout the project team.

When a motivation is appropriately used, as viewed by another, it is considered a strength. When the motivation is overused, as viewed by another, it is a weakness, which is to say that the other considers it counterproductive. Thus being analytic is a strength when analysis is called for, but being analytic is a weakness when it is time to harmonize or time to act. Similarly, being assertive-directing is a strength when direction is needed, but it is a weakness when the situation calls for, say, analysis. Being flexible is a strength when flexibility is needed; it is a weakness when steadfastness is needed.

People who are motivated differently tend either to admire or despise each other, rather than merely understand and accept one another, depending upon whether mutual

Table 6.3. Typical Ways That People Who Are Motivated Differently Regard Each Other

1. An Altruistic-Nurturing Person regards:

    An Assertive-Directing Person as fearless and vigorous[a] or as ruthless and domineering[a]
    An Analytic-Autonomizing Person as idealistic and objective[a] or as cold and stubborn[b]

2. An Assertive-Directing Person regards:

    An Altruistic-Nurturing Person as supportive and warm[a] or as sentimental and weak[b]
    An Analytic-Autonomizing Person as informed and analytical[a] or as impractical and nitpicking[b]

3. An Analytic-Autonomizing Person regards:

    An Altruistic-Nurturing Person as intuitive and understanding[a] or as irrational, sentimental, and subjective[b]
    An Assertive-Directing Person as enterprising and bold in planning[a] or as unorganized and impulsive[b]

[a]Positive attributes, when mutual respect and acceptance exist.
[b]Perceived negative attributes which may or may not actually exist, when there is disrespect, dislike, or suspicion.

trust, respect, and acceptance exist. Table 6.3 gives some common ways that people of one set of motivations perceive people of other motivations.

B. The Myers-Briggs Type Indicator®

Isabel Briggs Myers and Katharine Cook Briggs considered human behavior in terms of four preferences:

1. A preference for perceiving versus a preference for judging (or deciding). Those who prefer perceiving to judging would rather observe or study persons or situations than come to conclusions about them. And conversely.

2. A preference for thinking versus a preference for feeling, as a method of judging. Those who prefer thinking would rather arrive at a conclusion by analysis than by how a prospective outcome feels. And conversely.
3. A preference for sensing versus a preference for intuition, as a method of perceiving. Those who prefer sensing to intuition would rather focus on individual details than pay attention to overall patterns or relationships. And conversely.
4. A preference for considering the external world of facts and situations versus a preference for focusing upon one's inner thoughts, feelings, and ideas. Those who prefer dealing with external facts and situations are technically termed *extraverts*. Those who prefer dealing with inner thoughts, feelings, and ideas are technically termed *introverts*.

These preferences can be combined in sixteen ways, called *types*. Each type has certain corresponding strengths and weaknesses, some of which are significant in project work.

Before describing type characteristics, we note that individuals get pretty proficient at the things they prefer. That is, they enjoy what they prefer, so they do it more, so they get better at it. This does not mean that preferences absolutely dictate skills. Some individuals naturally have more skill in areas that they do not practice than others may have in areas that they do practice. However, individuals get better at things that they practice, which are generally related to their preferences.

Thus, preferences among project team members affect not only what occurs naturally in project work, but also the proficiency with which it occurs. As in the case of insight into the Porter model, insight into the Myers-Briggs Type Indicator will help project managers take advantage of opportunities and rectify shortcomings. For example, the following relationships pertain to problem resolution:

1. Intuitive-feeling types may be especially good at identifying problems.
2. Intuitive-thinking types may be especially good at identifying solutions.
3. Sensing-thinking types may be especially good at evaluating solutions.
4. Sensing-feeling types may be especially good at implementing solutions.

An implication of these relationships is that four types are needed for effective, efficient problem resolution. Or, conversely, if a group lacks certain combinations, it may not be as effective and efficient as necessary.

The complementary nature of opposites in the preference pairs is useful not only in resolving problems but more generally as well. The upper half of Table 6.4 shows how intuitive and sensing types can help each other, and the lower half shows how feeling and thinking types can do the same.

As in the Porter model, no one type is perfect. Each has a contribution to make, and a group with some of each is best. When a particular type is missing, the astute project manager will compensate for the lack in some structural way, such as calling in a local or outside consultant, holding formal or informal reviews, and using checklists.

Examples of using outsiders include the following:

1. If the group lacks an intuitive type, the project manager can have the project plan reviewed by an outsider before it is implemented in order to uncover unseen future possibilities.
2. If the group lacks a sensing type, the project manager can have the plan reviewed by an outsider in order to check the plan's ability to cope with immediate obstacles.
3. If the group lacks a feeling type, the manager can consult with one about how the customer may feel.
4. If the group lacks a thinking type, the project manager can ask one to help the group analyze its situation.

Table 6.4. Complementary Usefulness of Opposite Types

Intuitive types	Sensing types
Can raise new possibilites	Can raise pertinent facts
Can anticipate future trends	Can face current realities
Can apply insight to problems	Can apply experience to problems
Can see how different facts tie together	Can read fine print in a contract
Can focus on long-term goals	Can focus on immediate tasks
Can face possibilities with excitement	Can face difficulties realistcally
Can anticipate joys of the future	Can stay aware of joys of the present
Feeling types	**Thinking types**
Can forecast how others will feel	Can analyze consequences and implications
Can praise what is right and positive	Can find flaws in advance
Can make needed individual exceptions	Can hold consistently to a policy
Can teach and coach others	Can give needed critical feedback
Can stand firm for human-centered values	Can stand firm for important principles
Can organize people and tasks harmoniously	Can create rational systems
Can appreciate the thinker along with everyone else	Can be fair

Source: Modified and reproduced by special permission of the Publisher, Consulting Psychologists Press, Inc., Palo Alto, CA 94303 from *Introduction to Type®* by Isabel Briggs Myers.

Tables 6.5 through 6.8 list behaviors that tend to accompany each of the Myers-Briggs preferences. By comparing individuals' behavior with the four tables, their preferences can be inferred and their project behavior anticipated.

Each of these tables suggests that its two types behave oppositely. Indeed, often they do, sometimes conflictingly and sometimes to good advantage. Conflicts arise when one person believes that another should behave as the first does. But if opposites can see how they complement each other, then their differences can be very beneficial. Astute project managers minimize conflicts among opposite types on their project teams by helping them recognize and appreciate the

Table 6.5. Perceiving Preferences

Intuitive types	Sensing types
Like solving new complex problems	Like using experience and standard ways to solve problems
Enjoy learning new skills more than using them	Enjoy applying what they have already learned
Will follow their inspirations	May distrust and ignore their inspirations
May ignor or overlook facts	Seldom make errors of fact
Like to do things with an innovative bent	Like to do things with a practical bent
Like to present an overview of their work first	Like to present the details of their work first
Prefer change, sometimes, radical, to continuing what is	Prefer continuing what is, with fine tuning
Usually proceed in bursts of energy	Usually proceed step by step

Source: Modified and reproduced by special permission of the Publisher, Consulting Psychologists Press, Inc., Palo Alto, CA 94303 from *Introduction to Type®* by Isabel Briggs Myers. Copyright 1994 by Consulting Psychologists Press, Inc. All rights reserved. Further reproduction is prohibited without the Publisher's written consent.

Table 6.6. Judging Preferences

Feeling types	Thinking types
Use values to reach conclusions	Use logical analysis to reach conclusions
Want harmony and support among colleagues	Want mutual respect among colleagues
Enjoy pleasing people, even in unimportant things	May hurt people's feelings without knowing it
Often let decisions be influenced by their own and others' likes and dislikes	Tend to decide impersonally, sometimes paying insufficient attention to peoples' wishes
Tend to be sympathetic and dislike, even avoid, telling people unpleasant things	Tend to be firm-minded and can give criticism when appropriate
Look at the underlying values in situations	Look at the principles involved in situations
Feel rewarded when peoples' needs are met	Feel rewarded when a job is done well

Source: Modified and reproduced by special permission of the Publisher, Consulting Psychologists Press, Inc., Palo Alto, CA 94303 from *Introduction to Type®* by Isabel Briggs Myers. Copyright 1994 by Consulting Psychologists Press, Inc.

fact that together they can do better, more comprehensive and thorough work than either is likely to do alone.

The Myers-Briggs Type Indicator also indicates that introverts are likely to be misread. They are likely to conceal the perceptions or judgments that they prefer most and reveal their secondary preferences. Others are thus likely to believe that the second-priority values are first priority. This implies that introverts need to make an extra effort not to hide their primary feelings or thoughts and their perceptions, but rather to make them known.

Another result of the Myers-Briggs Type Indicator is that individuals who have some preferences in common tend to

Table 6.7. Perceiving Versus Judging Preferences

Perceiving types	Judging types
Enjoy flexibility in their work	Work best when they can plan their work and follow their plan
Like to leave things open for last minute changes	Like to get things settled and finished
May postpone unpleasant tasks that need to be done	May not notice new things that need to be done
Tend to be curious and welcome a new light on a thing, situation, or person	Tend to be satisfied once they reach a decision on a thing, situation, or person
Postpone decisions while searching for options	Reach closure by deciding quickly
Adapt well to changing situations and feel restricted without variety	Feel supported by structure and schedules
Focus on the process of a project	Focus on completion of a project

Source: Modified and reproduced by special permission of the Publisher, Consulting Psychologists Press, Inc., Palo Alto, CA 94303 from *Introduction to Type®* by Isabel Briggs Myers. Copyright 1994 by Consulting Psychologists Press, Inc. All rights reserved. Further reproduction is prohibited without the Publisher's written consent.

enjoy each other and work well together (unless they are competing with each other). However, a team's strength comes from having all perceiving and judging preferences present at least to some degree. Thus, the project team is better off with some overlap of preferences as well as some differences. While two people, one sensing-feeling and one intuitive-thinking, for example, have all the perceiving and judging traits together, they have little in common. An intuitive-feeling person and a sensing-thinking person are also needed if the group is have much harmony and succeed in its work.

Table 6.8. Concealing Versus Sharing Preferred Activities

Introverts	Extraverts
Like quiet for concentration	Like variety and action
Tend not to mind working on one project for a long time uninterruptedly	Often impatient with long, slow projects
Interested in the facts and ideas behind their work	Interested in the activities of their work and in how other people do it
Like to think a lot before acting, sometimes not acting	Often act quickly, sometimes without thinking
Develop ideas by reflection	Develop ideas by discussion
Like working alone without interruptions	Like having people around
Learn new taks by reading and reflecting	Learn new tasks by talking and doing

Source: Modified and reproduced by special permission of the Publisher, Consulting Psychologists Press, Inc., Palo Alto, CA 94303 from *Introduction to Type®* by Isabel Briggs Myers.

C. The Casse Communication Styles Model

Pierre Casse devised a model of human behavior based on orientations revealed by one's conversation and observable traits. Casse's four orientations are

1. *Action orientation—what*: results, objectives, achievements, doing
2. *Process orientation—how*: strategies, organization, situational facts
3. *People orientation—who*: relationships, teamwork, communications, feelings
4. *Idea orientation—why*: concepts, theories, innovation

Some people have one or two strong orientations and two or three weaker ones, while others seem to be more evenly balanced. Casse has developed a paper and pencil test that helps one to identify his or her orientation. One can often tell another's orientation simply by observing the person. Table 6.9 gives typical behavior and conversational content for individuals having strong orientations. When no single orientation or pair of orientations seems dominant, the person probably has a nearly balanced set.

With this background, we can apply Casse's model to project work. First, the model tells us that we can expect different people to rank action, processes, people, and ideas differently. Perhaps we have always known this, but many project managers seem not to acknowledge this fundamental principle or capitalize on it. Second, nearly every project needs all of these orientations at some time in its life cycle. Thus, responsibility assignments may have to be changed from time to time as the project evolves. Extremely action-oriented project managers can be a disaster during the project's conceptual stage or planning stage, while extremely idea-oriented individuals can be a serious handicap during the accomplishment stage of a by-the-numbers project. Extremely process-oriented individuals can hamstring a task that requires quick fixes, and extremely people-oriented individuals can cause unacceptable delays when conflicts arise if they put off decisions in order to achieve consensus.

As Porter said, the overuse of a trait is a weakness. Project managers have to curb the overuse of strong traits in Casse's sense as well as in Porter's and Myers-Briggs's.

D. Conclusions About Individual Motivation

Some general conclusions can be drawn from the related—yet different—Porter model, Myers-Briggs Type Indicator, and Casse model of behavior:

1. Project assignments should recognize that personality traits can be as important as technical skills in determin-

Table 6.9. Characteristic Conversational Content and Behaviors of Casse's Orientations

Conversational content	Behavior
Action orientation	
Results	Pragmatic
Objectives	Direct, to the point
Performance	Impatient
Productivity	Decisive
Efficiency	Quick
Moving ahead	Jumps from one idea to another
Responsibility	Energetic
Feedback	Challenges others
Challenges	
Experience	
Achievements	
Change	
Decisions	
Process orientation	
Organizing	Systematic
Procedures	Step-by-step
Planning	Logical
Facts	Factual
Controlling	Verbose
Analysis	Unemotional
Trying out	Cautious
Testing	
Observations	
Proof	
Details	
People orientation	
Self-development	Subjective
Feelings	Emotional
Awareness	Perceptive
Sensitivity	Sensitive

(*continued*)

Table 6.9. Continued

Conversational content	Behavior
	People orientation (cont.)
People	Spontaneous
Needs	Empathic
Motivations	Warm
Understanding	
Cooperation	
Communications	
Teamwork	
Team spirit	
Beliefs	
Values	
Expectations	
Relations	
	Idea orientation
Concepts	Imaginative
Creativity	Creative
Innovation	Full of ideas
Opportunities	Ego-centered
Possibilities	Unrealistic
Grand designs	Difficult to understand
Potential	Charismatic
Issues	Provocative
Problems	
Interdependence	
Alternatives	
New ways	
What's new in the field	

Source: Adapted from *Training for the Cross-Cultural Mind,* 2nd edition, by Pierre Casse. Copyright 1981, Intercultural Press, Yarmouth, ME. Used by permission.

ing success. Project managers should capitalize on individual strengths, compensate for shortcomings, and not try to depersonalize project work. Otherwise, they will miss the opportunity to rise above mediocrity.

2. Project managers should let individuals operate in ways that are comfortable for them, as long as these ways will work. Better commitment as well as better performance will result.
3. Project work changes as the project matures, and some reassignments may be in order. Project managers should plan for them in advance, so that they can be made gracefully and in time, rather than as belated recoveries from disasters.
4. Project teams should have all the needed traits available from the beginning, if possible. Project work is often unpredictable, and the project manager may not be able to rustle up the traits just when needed.
5. Every personality type is potentially of value. While sheer incompetence is not to be valued, project managers should not confuse personality and competence. Rather, project managers should harness competence by paying attention to personalities and arranging conditions to get maximum contributions.

This list might be expanded and the items refined or modified, but it is enough to make this point: personal factors are important in project management, and astute project managers learn how to use them to advantage.

7

Coordinating and Directing Techniques, Part II

This chapter focuses on communication and interaction aspects of coordinating and directing projects. While communications are not the province of project life alone, they are so important that project managers dare not leave them to chance; hence their inclusion in this book. *Interactions* as used here refers both to the way that project managers relate to the project team and to the arrangements needed for matrix organizations to work.

I. COMMUNICATION

Communication can occur in three ways:

1. Through words per se, as in memos and letters
2. Through vocals, such as tone, emphasis, hesitations, etc., as in telephone or face-to-face conversations
3. Through nonvocals or body language, such as facial expressions, gestures, position, etc., as in only face-to-face conversations

Generally, face-to-face conversations are richer in the amount of information transferred than telephone conversations, which are richer than written communications. This is particularly true when feelings are involved, which is often the case in project work. For example, a lack of belief, a commitment or a lack thereof, a defensive attitude, hopefulness, enthusiasm, and a host of other feelings that affect project performance can be detected when vocals and nonvocals are observed; they might not necessarily be detected from words alone. It behooves project managers, therefore, to communicate face-to-face and to urge their staff members to do likewise whenever serious issues are at stake.

It is also important for people who are working together for the first time to meet face-to-face and become acquainted. Then when they converse over the telephone they will have an idea of how the other's vocals might correspond to unseen nonvocals. This will enable them to read the other person a little more accurately.

To take good advantage of the available vocals and nonvocals, one must be an attentive listener, seeing as well as hearing what the other is transmitting. Unfortunately, some people are regularly poor listeners. Some reasons for poor listening follow, along with corresponding remedies.

1. They are so busy framing their replies that their thinking interferes with their seeing and hearing. If thoughts toward a reply do occur while they are listening, they should make a note or two for later reference and resume listening.
2. They feel dull and bored, unable to attend to each word. Bored listeners can sometimes help break the dull pace by asking appropriate questions.
3. They lack self-confidence and worry about the impression they are making, or are about to make, so that they do not pay attention to the speaker. This is most likely to happen in formal settings where each person knows

when he or she is due to speak. The best remedy for this is to practice in advance so that there is little doubt about one's ability to perform satisfactorily.

4. They dismiss the communicator as unimportant because he or she seems unable to hurt or help them. Cavalier dismissal of communicators can be dangerous. Perhaps these communicators will fade into the background and not volunteer information crucial to the project's success.

Sometimes one detects that others are listening poorly. In this event, it is appropriate to ask the poor listener a question that requires a response and wait for it. This can be done without putting anyone down and will usually bring all the listeners back into the conversation.

Good listeners also have some clear characteristics:

1. They can repeat back what has just been said. Beware, though, that false assurances may be gained if the nonvocals are missed.
2. They see the body language that accompanies the verbal message.
3. They detect minor discrepancies, small gaps or omissions, and minute ambiguities. Sometimes discrepancies, gaps, omissions, and ambiguities occur between different conversations. Questions can be asked to clarify the situation when they are noticed, unless a question would be distracting. Then it is better to probe later. But such little aberrations may tip off important information. What has happened, for example, to make the speaker more optimistic? Or pessimistic?
4. They notice when words, phrases, or expressions are used in an unusual or curious way, which may signify a critical group experience. A critical group experience may have been either very pleasant or very unpleasant. Until it is clear which type it is, one should avoid referring to it lest one upset a member of the group.

5. They are not embarrassed to ask for a moment to think when they have been too busy listening to frame a reply. One should feel free to ask for time to think before responding. It is courteous and shows that the listener has indeed been listening.
6. They habitually find value in everyone who seeks to communicate with them. Such listeners will find that they are among the first to know what is going on, rather than among the last.

II. CRITICISM

From time to time, a project manager will need to criticize the work of a team member, a supporting element, or even the boss in order to improve some performance. The art of giving criticism should be mastered carefully, for upon it hangs much of the project manager's ability to bring a project to a successful conclusion.

There is only one legitimate purpose of criticism, and that is to change behavior or performance. The presence of any other objective (e.g., to blame or shame another) is both inappropriate in a caring relationship and counterproductive. People simply do not feel like changing to accommodate someone who has just belittled them.

Since change is the objective of criticism, the recipient's ability to change must be gauged. Sometimes change is not possible: the desired change may lie beyond the other person's talent or knowledge, the person may be psychologically unable to change, or the desired change may be forestalled by matters outside his or her control. Whenever the recipient is unable to change, regardless of the reason, there is no point in giving criticism. In these cases, criticism only frustrates the recipient and does not produce the desired result.

Deciding whether another person is capable of changing is not necessarily easy, nor is one's judgment necessarily going to be correct. This does not mean that the would-be critic should play it safe and refrain from criticizing. Rather it means that one should try earnestly to assess the situation first and proceed gradually if one believes—but is not sure—that the recipient is able to change. By proceeding gradually, one can get some feedback before coming down too hard on the recipient if, in fact, the preliminary assessment is wrong. A caring disposition plus a normal amount of sensitivity to the other person will almost always enable a would-be critic to begin at an appropriate pace, read the signals given by the other, and proceed if on the right track.

Sometimes the recipient of criticism should be asked what would help him or her make the desired changes. Perhaps the situation can be negotiated, so that the critic receives what the critic truly needs and the person being criticized can perform in a way that he or she finds acceptable.

Criticism should be specific and in terms of the desired behavior or performance. If necessary, an example of acceptable behavior or performance should be provided. Generalities should be avoided, for they tend to divert the recipient's attention away from the desired change. Since a generality in criticism is usually accusatory, the recipient may search for an exception to disprove the accusation and argue over the past rather than focus on changing for the future.

The project manager is sometimes the recipient of criticism from one who is not very skillful. If so, he or she can help make the exchange productive by

1. Viewing the criticism as an opportunity to learn what is important to the critic
2. Drawing out the critic to obtain specific information on the changes desired
3. Watching nonverbal cues to see if the real substance of the critic's message is the same as the words
4. Not entering into needless arguments

III. MEETINGS AND CONFERENCES

An important vehicle for coordinating and directing project activities is the meeting. Most project personnel are familiar with meetings, but in their experience good meetings may be the exception rather than the rule. Successful project managers will correct any tendencies toward unsatisfactory meetings and make their meetings productive.

Successful meetings depend upon three factors:

1. Task management
2. Process
3. Physical arrangements

We discuss each of these factors below, as well as two factors that lie on the interface between task management and process, namely, agendas and minutes.

A. Task Management at Meetings

Task management refers to the business of the meeting, that is, the accomplishment of a specific task or set of tasks which is the purpose of the meeting. Accordingly, task management begins when one decides on the goals of the meeting.

Potential meeting goals are to

1. Define a problem
2. Create alternative solutions to a problem
3. Weigh courses of action
4. Decide among alternatives
5. Plan future activities
6. Exchange information
7. Review and clarify, i.e., to make sure everyone understands
8. Evaluate progress
9. Raise issues

Sometimes it is appropriate to pursue just one of these goals at a single meeting. At other times it may be appropriate to pursue two or more, such as goals 1 through 4 or goals 6 and 8.

Project managers and other meeting attendees should realize that some participants may have hidden meeting objectives which may interfere with accomplishing the stated goal. Typical hidden objectives are to

1. Block, delay, or confuse action. Some people will use a meeting to muddy the waters because they do not like the course being followed.
2. Fill up time, avoid other work, or get a free lunch.
3. Fill a meeting quota. Some meetings are scheduled to occur periodically whether they are needed or not. However, this arrangement may be valid if it is important for the participants to touch base regularly.
4. Keep the group divided. When combatants are brought together, they are likely to harden their positions so that compromise cannot be achieved.
5. Diffuse decision responsibility so that it will be hard to blame anyone in particular for a bad decision.
6. Meet the social needs of the group. This is a legitimate purpose but it should not be confused with efforts to accomplish anything else.
7. Use the meeting as an arena to impress the customer or boss.

Some of the normally hidden objectives can be a legitimate meeting goal at one time or another. However, difficulties arise when different individuals have conflicting goals or when an individual's objective is at odds or cross purposes with the primary purpose of the meeting. In these cases, hidden objectives should be smoked out and confronted. The meeting goals that will be pursued will have to be decided and agreed upon before any useful work can be done.

Related to the goal of the meeting is the scope of the meeting. Scope refers to the extent to which the subjects will be explored or developed. It makes a difference, for example, whether a meeting to develop budgets will be concerned with budgets to the nearest hour of effort or to the nearest $10,000. Likewise, it makes a difference whether a meeting to consider oil drilling is to be concerned with all of North America or just the off-shore region of California.

Once the purpose and scope of the meeting are determined, it is easy to decide who should be invited. Conversely, if the purpose is not known, it is impossible to decide whom to invite. The list of invitees should include those who can contribute toward the meeting's goals, and only those people. This does not mean that those who have little to say on the subject but need to hear should be excluded; indeed, they should be included. But it does mean that the meeting should not include idle observers who do not need to hear the proceedings. Their presence merely dilutes the attention of the real participants and it may also inhibit their candor.

Having determined the goal, scope, and participants for the meeting, the leader and the participants should determine what homework they must do in order to perform their individual roles at the meeting. The invitees should understand the goal and scope well enough that they can easily determine their homework.

Knowing the meeting's purpose, scope, participants, and the participants' needs for preparation, the chair can then schedule the meeting. The meeting must be held soon enough for its result to be timely and late enough to allow the participants to prepare. It must also be scheduled so that the group has enough time to do the work that has to be done.

If at all possible, the length of the meeting should be set in advance and the participants notified. This allows them to schedule their own time better. If it is useful to keep the meeting from running on and on, then it can be scheduled shortly before lunch or quitting time. Most people will conclude an otherwise interminable discussion in order to eat or go home.

Finally, the location of the meeting must be decided. Both the purpose of the meeting and convenience to the participants are factors. Sometimes the purpose dictates or influences the choice of meeting location. Perhaps it would be useful to meet at the job site, perhaps in the vicinity of files or other resources that might be needed. Perhaps there is a psychological advantage to be gained (or lost) by one of the parties if the meeting is held in a particular location. Perhaps the meeting should be held at a neutral location. Perhaps the meeting should be held away from possible interruptions and distractions. Whatever location is selected, it should be chosen consciously with due regard to the impact that it can have on the success of the meeting.

B. Meeting Process

Meeting process refers to the way the participants behave and interact at the meeting. A meeting's objectives are not met simply because the attendees assemble in a room. Rather, certain activities must occur, such as initiating the discussion, obtaining information, and achieving agreement, if there is to be any progress toward the goal.

Key process activities are

1. *Initiating.* The chair is responsible for starting the meeting and initiating discussion, but all of the participants can raise issues or introduce new topics when appropriate, and they should raise important items that are being overlooked.
2. *Information and opinion providing.* All meeting participants have a responsibility to share relevant information and opinions, whether or not they have been directly asked. After all, they were invited to the meeting for the contribution they can make toward its goals.

3. *Information and opinion seeking.* In some meetings, not all the participants will volunteer what they know or think or feel. Those who are quiet should be drawn out. Some may even have to be coaxed.
4. *Encouraging.* Venturing an opinion, offering information, or just asking a question represents a psychological risk for some meeting participants. These people need to be encouraged, and everyone at the meeting should present an encouraging rather than a discouraging attitude.
5. *Reality checking.* Assumptions, opinions, and speculations offered in the meeting need to be checked against facts. Whoever has factual knowledge about the subject being discussed has a responsibility to verify or challenge assumptions, opinions, and speculations.
6. *Analyzing.* From time to time it will be appropriate to analyze what a situation really is, what some information really means, what is known and what is unknown, etc. Participants can contribute greatly to the progress of their meetings if they will do appropriate analysis and share the results with their fellow attendees.
7. *Clarifying.* It is occasionally necessary to make sure that everyone understands—and has the same understanding. When there are expressions of bewilderment or confusion, the participants should clarify the issues before proceeding.
8. *Consensus testing.* As a subject is developed in a meeting, it is important to determine if the participants agree or not on its various aspects. Formal votes are usually not required. If one cannot tell from the participants' comments or facial expressions, a simple question such as, "Do we all agree on this?" may be all that is needed.
9. *Harmonizing.* Points of agreement should be searched for and disagreements should be harmonized or re-

solved wherever possible. Whoever sees a way to reduce a disagreement should help do so by identifying points of similarity, alternative interpretations, possible compromises, and so forth.

10. *Summarizing.* It is important to summarize a discussion after it has been underway for some time, or just before moving to a new topic, even if the discussion was not particularly long. This provides the entire group a compact statement of the key issues and identifies any loose ends that might exist. The group can then move forward with a sense of having reached a milestone. Also, the participants will have a common statement of what remains to be done, if anything.
11. *Recording.* Meeting discussions and gains made should not be lost or subject to faulty recall in the future. Therefore, the chair should see that at least one person takes notes of the meeting. Alternatively, major items can be written on a flip chart or chalkboard for all to see and then later transcribed. Meeting records are discussed further in Section III.C.2.
12. *Goal tending.* Goal tending refers to keeping the discussion on the topic and curtailing useless or untimely excursions. All meeting participants have a duty to tend the goal and to retrieve and refocus the discussion if it has strayed afield.
13. *Time keeping.* Time keeping means avoiding unnecessarily protracted discussions and moving forward from one agenda item to the next as quickly as practical. As in the case of goal tending, all meeting participants have a duty not to waste meeting time and to help others avoid wasting meeting time.
14. *Gate keeping.* Gate keeping means assuring that one or a few individuals do not dominate the discussion and thereby exclude others from participating. Again, participants must practice self-discipline so that they do not take up more than their fair share of the meeting time.

They also have a duty to intercede if another attendee does not practice self-restraint.

15. *Listening and observing.* If a meeting is not going well, whoever notices that something is amiss should listen to and observe the process in order to identify what is missing. He or she should then supply the missing element if possible or encourage others to provide it.

Some behaviors are productive at meetings and others are counterproductive. Productive behaviors include the following:

1. *Participating.* People are invited to a meeting in order to participate in the discussion. If they are there just to listen, they might as well not come, generally speaking. They can read the minutes instead.
2. *Sharing strong feelings.* When people say quietly to themselves or friends after the meeting, "I wish I had said . . . ," they probably should have said it in the meeting. If they had, they might have found that they had company. Perhaps their comments would have changed the course of the discussion. Or they might discover that they are alone in their feelings and then wish to revise their positions.
3. *Using good timing.* One should make reminder notes of ideas that properly belong later. Similarly, one should not wait until the end of the discussion to say something that should have been said earlier.
4. *Making sequence responses.* A disjointed discussion is difficult to follow. To minimize this problem, participants should provide transitions between ideas already discussed and what they are about to say. In other words, participants should make apparent or obvious the sequential relations that exist between what they will say and what has gone before.
5. *Building on ideas.* Other things being equal, it is more economical of meeting time if ideas already discussed are

expanded, embellished, enhanced, and made better rather than displaced by new ideas. Improving and supporting another's idea helps consolidate the participants and move them toward agreement, while needlessly introducing new alternatives may divide the group and make agreement more difficult to reach.

6. *Paraphrasing to show understanding.* Just as in conversations, it is helpful to check one's understanding by paraphrasing what one believes to be the case.
7. *Giving nonverbal as well as verbal cues.* Facial expressions and other nonverbal cues help participants know if they are communicating clearly, have support for their ideas, are boring their listeners, and so forth, without having to poll the assembly. This enables speakers to adjust their comments to the listeners' needs and expedite the flow of the meeting.
8. *Having courage and taking appropriate risks.* It takes courage at times to express views that may not be popular, to question another's assertion, to ask for time to analyze the situation to bring another back to the subject, etc. Yet such courageous acts may be keys to advancing the progress of the meeting. Meeting participants should have courage and take appropriate risks in expressing themselves and in dealing with other participants.

Counter-productive behaviors in meetings include the following acts

1. *Ego-tripping,* which is at the expense of everyone else's time
2. *Inappropriate entertaining,* i.e., entertaining which disrupts the meeting and does not contribute toward either the task or process of the meeting
3. *Subgrouping,* where a little pocket of people holds its own meeting, both withdrawing from the main meeting and disrupting it at the same time
4. *Withdrawing* mentally from the meeting

5. *Blocking,* i.e., obstructing the flow of the meeting, being argumentative for its own sake, and so forth
6. *Dominating,* which not only takes up other people's time but prevents them from making their contributions
7. *Sidetracking or diverting* the meeting from its purposes

As noted throughout this section, meeting process involves all the attendees. Each one is responsible for the success of the meeting. Each one should perceive what is going on in the group, identify what is missing in either process or task, and supply it if possible or see that it is supplied.

The meeting chair has additional duties which are not shared with the rest of the participants: the chair sets the stage for the discussion and introduces the meeting. The chair also introduces the participants to each other when they have not met before.

At the end of the meeting, the chair should thank the participants. The chair needs also to be sure that the participants receive a list of action assignments. Each entry in the list should include the name or description of the assignment, the assignee, the date due, the form of any deliverables, and the recipients of deliverables. If meeting notes or minutes are to be circulated (see Section III.C), the chair also should ensure that this occurs in a timely way. The assignment list may then be incorporated in the minutes or appended to them.

C. Agendas and Minutes

Agendas and minutes lie at the interface between task management and meeting process. They are connected at once to both the purpose of the meeting and the conduct of the meeting.

1. Agendas

In simple terms, an agenda is a list of things to be done. Often printed agendas, especially those distributed publicly, are merely lists of the topics to be considered at the meeting. A meaningful agenda, however, shows not only the topic or name of the issue but also its goal and scope, the key participants, and some notion of the amount of time to be devoted to it. Meetings often flounder without an agenda, and it is generally wasteful to proceed until an agenda is agreed upon. The chair is responsible for drafting an agenda.

It is useful for the participants to review the chair's proposed agenda and modify or revise it as appropriate as their first item of business after introductions. This approach has three benefits. First, it may result in a sequence that makes better use of everyone's (or at least most people's) time than the original agenda. Second, it helps the participants understand each other's priorities and interests. And third, it gives the participants a chance to trade small concessions and begin to build good will toward each other (see Chapter 9).

2. Minutes

Minutes, that is, a record of the important aspects of a meeting, should be made of every meeting. Their form and formality may vary, depending on the circumstances.

All minutes should include the following information:

1. The time and place of the meeting
2. The attendees, including part-time attendees and guests
3. A copy of the agenda (as an attachment)
4. A brief preamble on the purpose of the meeting
5. Notes on issues discussed, including concerns, assumptions, decisions made, and open issues
6. Action assignments (see Section III.B)
7. When and where the next meeting will be, if known

If formal minutes are called for, they should be organized, edited for clarity and accuracy, typed, and distributed within a few days of the meeting. The attendees should receive the minutes while their memories are still fresh. This will enable them to review the minutes meaningfully and identify any errors of omission or commission. If minutes are received after memories have dimmed or become selective, needless and perhaps unresolvable arguments can develop about what really transpired at the meeting.

Advantage accrues to the one who prepares the minutes. There is usually more than one way to express what happened at the meeting, and the individual who prepares the minutes can choose the words. In time (i.e., after the minutes are formally approved if that practice is followed, or after no one complains if that is the practice), the minute taker's record will be the only official record. Thus, those who refer to the official record later will have his or her version of the proceedings.

D. Physical Arrangements for Meetings

The success of the meeting may depend on its physical arrangements. While satisfactory arrangements cannot guarantee success, unsatisfactory arrangements can almost certainly assure a more tiring and less productive meeting.

The following conditions are essential for good meetings:

1. Comfortable air quality and temperature
2. Ability of the participants to see and hear each other without straining
3. Freedom of the participants to stretch, get up, move about, etc., without disturbing others
4. Soft but adequate lighting and no glare from windows, walls, white boards, etc.
5. Proper furniture

IV. FACE-TO-FACE AND TELEPHONE CONVERSATIONS

Not all project interactions require or indeed are well served by meetings. Face-to-face and telephone conversations on one hand and written communications on the other are alternative ways to conduct project business. Conversations are discussed in this section and written communications are discussed in the next. In both cases, attention is focused on the purposes that each can serve and on ways to make them effective.

A. Purposes of Conversations

Conversations are two-way exchanges and are best used where it is important to know how well one is communicating while proceeding or where one of the parties needs some response before proceeding. Thus, conversations can be effectively used for the following purposes:

1. To give information and instructions
2. To obtain information
3. To coach
4. To seek cooperation
5. To keep in touch

Information and instructions can be given by written communication, as discussed below, but there are times when a conversation is better. Whenever the information or instruction is subject to partial or alternative interpretation, depending on the background and understanding of the recipient, it is helpful to communicate orally. This way, the sender can check periodically to verify that the receiver has grasped the information or instructions as they were intended. The sender can observe vocals and nonvocals as well as ask the receiver to paraphrase to ascertain understanding. Once understanding has been achieved, it can be confirmed in writing for the record if necessary.

Likewise, information can sometimes be obtained most readily through a conversation. This is particularly true where the request is subject to being misinterpreted or where the scope needs to be developed by the requester and the responder together before the request can be addressed.

Again, the request can be confirmed in writing once it has been adequately developed through conversation.

Coaching consists of giving instructions one at a time where each instruction is tailored to the recipient in light of the recipient's response to the previous instructions. Coaching cannot be effectively done in written form.

Cooperation is also better developed by conversation rather than written communication. When seeking cooperation, one needs to be aware of the other person's responses in order to know best how to unfold the request. Also, the person whose cooperation is being sought generally finds it more difficult to ignore a face-to-face or telephone request than a written request. Moreover, a conversational request affords an opportunity to negotiate the exact form or content of the cooperation. For all these reasons, a conversational request is more likely to elicit cooperation than a written request.

Keeping in touch is an important part of project management, whether the person to be kept in touch with is the customer, task leaders, support function personnel, subcontractors and vendors, or the boss. Project managers typically can use as much insight as they can get into other peoples' intentions and concerns. This insight will help them prepare for possible changes in goals or plans. If they wait until the changes have such a high status that they are written down, then they may find it difficult to accommodate them. On the other hand, if they get early indications of possible changes, then they can prepare for them with less difficulty. Keeping in touch establishes and maintains an open communication channel for the parties to exchange ideas without having to organize their thoughts into self-consistent expositions and without committing the thoughts to writing. People will often share notions with others if they can do so effortlessly and without worrying whether they will be held to them. Conversation promotes such sharing, while writing inhibits it.

B. Preparing for a Conversation

One should prepare for all but impromptu conversations the same way one prepares for a group meeting. One needs to know the goals and the scope of the discussion and needs to do whatever homework may be necessary. Also, one needs to consider how to begin the discussion so as to encourage the other person to continue the conversation rather than end it.

C. Conducting the Conversation

A productive conversation consists of several distinct steps. Often they seem to occur effortlessly. However, no one should be deluded into thinking that they will occur automatically. Some attention must be paid to conducting the conversation by at least one of the parties to the conversation. Project managers will enhance the conversations that they initiate if they take responsibility for the following steps:

1. *Establishing the purpose of the conversation.* The project manager should let the other person know the ostensible reason for the conversation. One should not have to wonder about or guess the purpose. It is true, however, that the project manager may wish to reveal only part of the purpose at the outset. The project manager may want to wait until he or she has some insight into the other person's viewpoint or frame of mind before revealing the rest of the purpose. If an invitation is extended or an appointment is made to hold the conversation in advance of the conversation itself, it is appropriate to convey the purpose and scope of the conversation then. This practice will enable the other person to prepare for the conversation if necessary.
2. *Obtaining the other party's interest.* Directly related to establishing the purpose of the conversation is obtaining the other party's interest. The other party must feel that

it is more worthwhile to hold the conversation than to do anything else at this time. Unless the purpose itself makes the significance of the conversation patently clear, it is probably necessary to explain why the topic is important enough to warrant the other party's time and attention.

If others have shown preferences for action, processes, people, or ideas, along the lines of Casse's communication styles described in Chapter 6, then catering to their preferences will help capture their interest. Table 7.1 lists ways to enhance communication with parties with these four preferences.

3. *Exchanging information.* Both parties are more likely to consider the conversation productive if they exchange information in small increments. First, this practice allows each to determine how much the other knows about the subject so that his own comments neither assume too much nor are needless. Second, it allows each to partici-

Table 7.1. Communicating with Another Person

With an *action-oriented person*: focus on results; emphasize practicality; be brief; have a backup approach but do not dwell on it; use visual aids.

With a *process-oriented person*: stick to the facts; organize your material into a logical order; be precise; include options.

With a *people-oriented person*: allow for small talk; emphasize relationships between your proposal and the people concerned; show how the idea worked well in the past; indicate support from others; be informal.

With an *idea-oriented person*: emphasise key concepts that underlie your proposal; work from the general to the particular; allow sufficient time for discussion, including tangents; emphasize uniqueness and the future value of your approach.

Source: Adapted from *Training for the Cross-Cultural Mind*, 2nd edition, by Pierre Casse. Copyright 1981, Intercultural Press, Yarmouth, ME. Used by permission.

pate actively in the conversation so that each one's interest is maintained. And, third, it allows each to tailor what one says to the other's mood and thereby avoid both creating a bombshell and missing the target.

4. *Assuring understanding.* Each party in a conversation needs to be understood and to know that he or she is understood. If clear and complete understanding is not evident from the other's remarks, one should ask the other to confirm understanding by paraphrasing what has been said. Likewise, one can volunteer to paraphrase to show one's own understanding. If a difference of understanding appears, the issue should be explored or developed further until the two parties agree. Whatever disagreements remain should be about preferences, opinions, or values and not about facts or what transpired in the conversation.
5. *Obtaining commitments and establishing follow-through assignments.* When one party to the conversation wants the other to do something, he or she must obtain the other's commitment to do it. Many times either or both parties will be ready to agree to do something but the commitment or agreement itself is not vocalized. It is important, however, for these agreements to be made explicit. Both parties should state unequivocally what they are going to do and by when they will do it. It is often a good idea for one of the parties to confirm their agreements in writing. And who shall write the confirming memo shall be one of the items agreed upon.
6. *Recording salient details.* Wise project managers will always prepare their own records of salient conversational details. They may or may not provide a copy to the other party. These records should include not only information on what transpired but also notes on lingering concerns, possible threats and opportunities, and other "intelligence" that might later be useful.

V. WRITTEN COMMUNICATIONS

Written communications are an alternative to meetings and face-to-face and telephone communications. Like the other forms of communication, they are better in some applications than others. As in the previous section, we discuss the purposes that written communications serve well and ways to use them effectively.

A. Purposes of Written Communications

Written communications are most useful when precise expression is important, when a dialogue is not needed, or when the recipients are so dispersed that oral communication is quite impractical. They should not be relied upon to coach, to elicit cooperation, or to keep in touch. Thus, written communications can be used effectively for the following purposes:

1. To give or request information and instructions that do not involve interactions between the parties
2. To issue reminders
3. To confirm, record, or document earlier conversations

B. Style

Style is an important aspect of effective writing. The ultimate measure of style is whether or not it helps the reader understand the writer's message. Styles that detract from understanding are bad and styles that help are good. Thus, the preferred style depends partly on the readers and their ability to understand from the style used. Nevertheless, a general axiom or two will serve all writers well.

Perhaps the most important advice is to put first things first. The subject and its importance to the readers should be stated at the outset. Project personnel are busy and will not wade through a memorandum or letter to determine if they should have read it. Unlike a conversation, the writer has only one chance to capture the reader's interest. If one does not seize the reader's attention immediately, one might as well not send the communication at all.

Also, the more important points of the communication should be presented before the less important points. If the reader should get only part way through the communication, what he or she has already read should be more important than what he or she will miss by not reading the rest.* Moreover, truly significant elements should be highlighted and not buried in continuous prose.

The writing should also be clear and simple. Long sentences, perhaps composed of subordinate clause upon subordinate clause, should be avoided. Try to communicate to your readers rather than impress them. Where possible, use a phrase instead of a clause and a word instead of a phrase. While keeping to simple sentences, avoid monotony. Read your writing aloud to see if it flows without being a torrent.

C. Content of Written Communications

Written project communications should contain what the readers need to know and only that. A memorandum or letter should not be cluttered with extraneous material. (Readers should also watch for what might have been said but was omitted. There may be significance in the omission.)

A written communication should also tell its readers what the information means and why it is important. This should not be left to their imagination.

*Suggesting that the most important material be placed first assumes that all potential readers have similar priorities. If they do not, then they may need different versions or at least a guide in the introduction to tell them where they will find what most interests them.

Finally, the communication should tell the readers what action, if any, they should take and when. It should also tell them if a response is expected.

D. Memo and Letter Organization

There are many possible ways to organize a lengthy, complex memorandum or letter. The following list can be used as a suggestion menu if one has trouble identifying a possible approach. Sometimes two or more ways can be combined. One way, say, geographical, can be the primary organizing framework and another, say, chronological, can be an organizing framework within each primary element.

1. Chronological
2. Geographical, e.g., site A, site B
3. From the known to the unknown
4. From the concrete to the abstract
5. From the simple to the complex
6. From the general to the specific
7. Cause and effect
8. Problem description, analysis, and solution
9. Functional, e.g., by discipline
10. From the most important to the least important
11. Questions and answers

Numbers 3 through 6, depending on the circumstances, can be especially helpful when readers need to understand a complex situation. In the case of number 8, the solution can be put first if readers already have some idea of the problem.

VI. THE FIEDLER CONTINGENCY MODEL OF TEAM EFFECTIVENESS

The Fiedler contingency model of team effectiveness is based on the idea that group performance depends upon four factors, two of which concern interactions between the team leader and team members. The four factors are:

1. Whether or not the leader enjoys good relations with the team members
2. Whether the task is highly or poorly structured
3. Whether the leader has a high or low position of authority and power
4. Whether the leader displays a task-orientation or a relationship-orientation toward the group's effort.

Leader–member relations are described as good if the people look forward to working with each other and there is mutual respect and trust. They are described as poor when there is a lack of respect or trust and some people would rather not work with the leader.

A task is said to have high structure when both the goal and methods are well understood and agreed upon and there is little ambiguity about individual roles and responsibilities. A task is poorly structured when any of these ingredients is missing.

A leader's authority and power in the eyes of project team members can arise from having a high position in the organization, from reporting to a very high level on the particular assignment, or from having personal expertise, independent of organizational factors.

Task-orientation refers to focusing on goals, procedures, and assignments, with little attention to how the team members interact. Relationship-orientation, on the other hand, refers to focusing on how the team members behave, with less direct attention to specific goals, procedures, and assignments. Various combinations of these four variables are shown in Table 7.2, together with Fiedler's prediction of how favorable the situation is for the leader and how successful the leader is likely to be. These combinations correspond to the eight numbered categories in the table.

By assuming that a leader can be truly successful only when the team is successful, the table can also be read to

Table 7.2. Fiedler's Contingency Model of Team Effectiveness

Situation characteristics								
Leader–member relations	Good				Poor			
Task structure	High		Low		High		Low	
Leader position power	Strong	Weak	Strong	Weak	Strong	Weak	Strong	Weak
Situation designation	1	2	3	4	5	6	7	8
Situation rating[a]	Favorable			Moderately favorable		Unfavorable		
Probable leader performance								
Relation-oriented leader	Poor			Good		Good?	??	Poor
Task-oriented leader	Good			Poor		Poor?	??	Good

[a]Favorableness of the situation for the leader and, by implication, for the team.

determine whether a task orientation or a relationship orientation is more likely to enhance project success. In general, a task orientation is more likely to enhance project success when the situation is classified as Category 1, 2, 3, or 8, and a relationship orientation is more likely to be helpful when the situation is classified as Category 4 or 5 and possibly 6. Whether a task orientation or a relationship orientation is better for Category 7 is unknown.

If a relationship-oriented leader converts a situation initially classified as Category 5 into a Category 1 situation or a Category 6 situation into a Category 2 situation, then the leader should adopt a task orientation in order to maximize the outcome. Similarly, if a relationship-oriented leader's team converts a Category 4 situation into a Category 2 situation, the leader should change to a task orientation.

A few words of caution are in order before leaving this discussion of Fiedler's contingency model. First, it is one of several contingency theories which, in general, are concerned with group performance as a function of the situation. Fiedler's model was selected because it seems to relate as well as any to situations like those of a project manager. Second, it is very difficult to replicate study populations and to control experimental conditions in order to validate contingency models. Accordingly, the validity of Fiedler's model, as well as other contingency models, is somewhat controversial.

VII. MATRIX ORGANIZATIONS

Projects increasingly occur in situations (1) where critical personnel resources must be shared, because of scarcity or the need to provide full-time work; (2) where personnel assignments must be rearranged without reorganizing departments; (3) where specialists must keep in touch with each other, in order to share ideas and keep current in their specialty; or (4) where administrative functions such as task assignments and salary administration must be performed concertedly for a group of specialists who work on different tasks. When these conditions are met by *some* people having more than one boss or project customer, a *matrix structure* exists. If this structure is supported by mechanisms, a culture, and behavior that enhance the effectiveness and efficiency of these people, then a *matrix organization* exists.

Compared to most other organizational arrangements, matrix arrangements require greater interpersonal skill and involve more advanced planning, coordination, and negotiation, as described below. Organizations adopting matrix arrangements may also have to relax formalities in order to take advantage of resource sharing and to change their reward structure to favor cooperation rather than competition. Despite these factors, however, matrix arrangements are often used when critical resources must be shared among multiple tasks and organizational flexibility is prized.

This section discusses the structure of matrix organizations and planning and coordinating mechanisms and behaviors that are needed for a successful matrix organization. Cultural aspects of successful matrix organizations, which are typically beyond the control of individuals (except, perhaps, the highest boss in the parent organization) are discussed in Appendix F.

A. Matrix Structures

The essential structural feature of a matrix organization is that some people have two or more bosses or project customers. People with more than one boss are called *multibossed individuals,* and their bosses and customers are called *multiple bosses.* Somewhere in the organizational hierarchy above the multiple bosses is a *common boss,* who resolves conflicts between multiple bosses when the latter are unable to do so themselves. These roles are depicted in Figures 7.1 through 7.4. A project manager could be any of the three positions, depending on local circumstances.

Figure 7.1 shows the simplest form of matrix structure, where the multibossed individual has just two multiple bosses, and the common boss is directly above the multiple bosses. The triangle at the bottom represents people who work solely for the multibossed individual. Figure 7.2 shows a slight variation, where the common boss is separated from the two multiple bosses by intermediate bosses; the number of intermediate bosses does not have to be the same on both sides.

A matrix arrangement may have more than one common boss, depending on the issues involved. Thus in Figure 7.3 the project managers and the group supervisor are multiple bosses, and the common boss is either a program director or the department manager, depending on the breadth of the issue. If the issue concerns priorities among projects within a single program, for example, then the common boss is the program director. On the other hand, if the issue involves two or more programs or involves responsibilities of a group or section as well a program, then the common boss is the department manager. In Figure 7.4, if the issue involves the interests of more than one department, the common boss is the division vice president.

Matrix arrangements allow individuals to have different roles at the same time. Thus, a group supervisor may concurrently be a multiple boss (as in Figure 7.3) with respect to some assignments and multibossed individual (as in Figure 7.4) with respect to other assignments. Likewise, a project manager may concurrently be a multibossed individual if working on projects for more than one program director and a multiple boss with respect to the individuals supporting these projects.

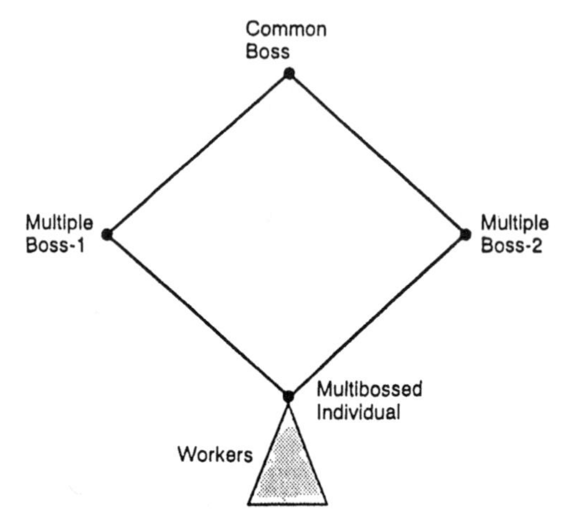

Figure 7.1. The simplest form of matrix structure

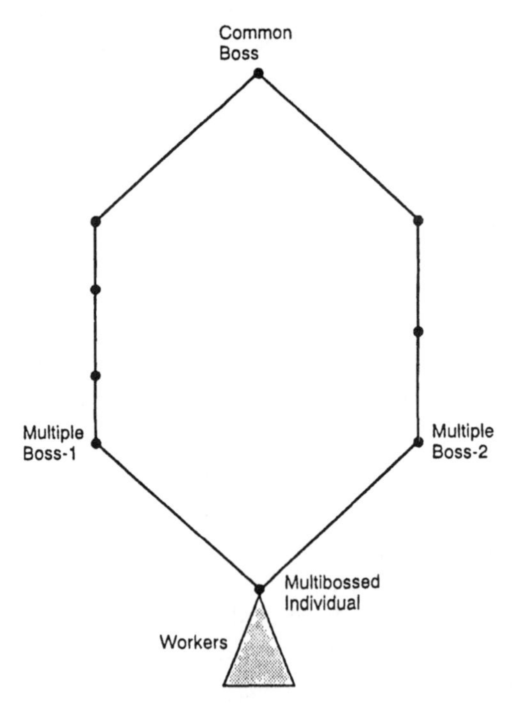

Figure 7.2. A matrix structure with several organizational levels between the multibossed individual and the common boss.

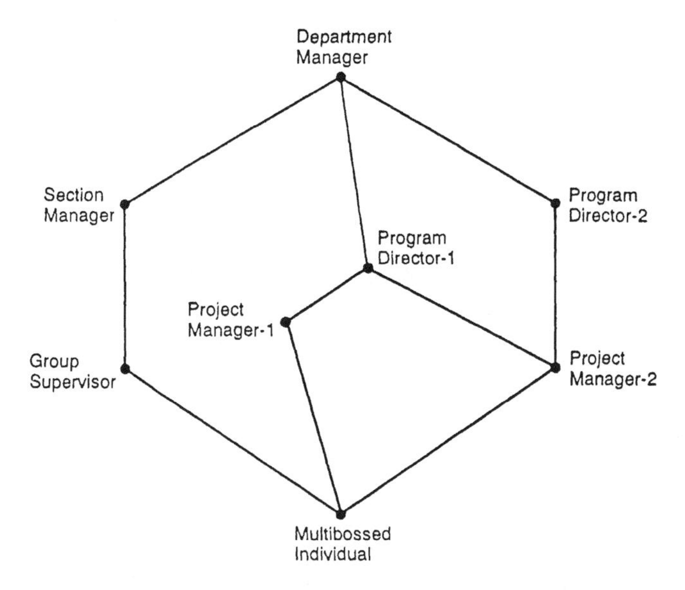

Figure 7.3. A somewhat complex intradepartmental matrix structure.

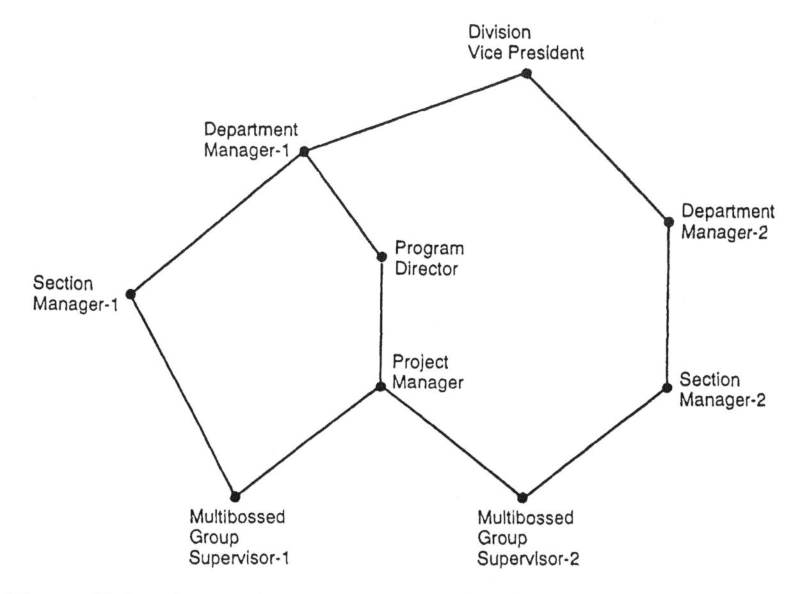

Figure 7.4. A matrix structure involving two departments.

B. Matrix Planning and Coordinating Mechanisms

In addition to structural aspects, successful matrix arrangements involve planning and coordinating mechanisms focused on the availability of multibossed individuals. Guidelines to aid in these activities are as follows:

1. *What should be planned?* From a matrix standpoint, the critical part of planning is scheduling assignments of multibossed individuals. Plans should be sufficiently detailed that opportunities to optimize assignments of these personnel can be detected. At the same time, multibossed individuals should not be so tightly scheduled that mere hiccups in their work for one multiple boss create undue problems for other multiple bosses relying on the same individual.
2. *Who should organize planning activities?* Common bosses should make clear how much initiative they will take to coordinate their multiple bosses' plans *versus* how much they are leaving to the multiple bosses. Unless the number of multiple bosses is small, (say, three or fewer) *and* each multiple boss shares only one or two resources, the common boss may have to personally organize and supervise the planning efforts. Otherwise, the sheer number of boss-resource combinations is more likely to lead to chaos than to a set of mutually consistent plans.
3. *Who should participate in planning?* Common bosses should ensure that planning is done at least by those who will carry out the plans, in order to tap their expertise, to gain their commitment, and most importantly in a matrix organization to have them identify potential conflicts that must be resolved before they occur. Multibossed individuals working for a variety of multiple bosses are typically best able to spot potential conflicts and have a special duty to identify them.

   Some multiple bosses tend naturally to plan without involving their multibossed individuals. Common bosses can forestall this tendency by approving only those plans that have appropriate participation by multibossed individuals.

Multiple bosses should inform each other of their own thinking, for often they are the actual or proximate source of changes that require planning and coordination. Equally important, multiple bosses should consult their multibossed individuals early to identify potential conflicts, work to resolve them, and make their own commitments in light of these resolutions. If multiple bosses cannot resolve conflicts by negotiating among themselves, then they should seek resolution from their common boss.

4. *How detailed should plans be?* A main purpose of planning is to find and solve problems in advance. Thus, plans for matrix arrangements should be sufficiently detailed to expose conflicts in assigning multibossed individuals and to facilitate conflict resolution. Often, this means that plan increments should be at the level of days or half-days rather than weeks or months.
5. *What information should be exchanged once plans have been prepared and are being pursued?* Plans are estimates of what will happen and need to be revised to accommodate realities as work progresses. These revisions need to be coordinated when multibossed individuals are involved.

   A multiple boss should not assume that the plans of other multiple bosses are either completely frozen or completely flexible. Rather, each multiple boss needs to openly describe how conditions have changed and what options are possible, acceptable, and preferred.

   Options need to be described in sufficient detail that fine adjustments can be made, and multibossed individuals may recognize ways to ameliorate conflicts in their multiple bosses' plans by rearranging their own assignments. The costs and benefits of the options need to be understood and win–win solutions developed. The overarching principle should be the greatest good for the greatest number, with no truly intolerable adverse consequences.

6. *When should plans be prepared and revised?* Planning that might involve multibossed individuals should be started as soon as possible, even before the work is authorized. The aim is to identify potential conflicts and solutions so that work already authorized can be coordinated with proposed work and not foreclose options that may be needed if the proposed work indeed materializes. Thus, planning should be ongoing and concern prospective as well as existing work.

   Plans should be revised periodically, neither so often as to cause organizational fibrillation (i.e., frequent but unproductive readjustments), nor so infrequently as to not resolve intertask resource conflicts in a timely way.

   To minimize organizational fibrillation, *changes* of priorities and resource assignments can be batched and acted upon at appropriate intervals, which can be set by common bosses. Similarly common bosses should restrain themselves in order to avoid causing organizational fibrillation.

7. *What atmosphere and pace are needed for planning and coordinating?* Planning in a matrix organization requires personnel to share their opinions and concerns as well as plain information. Only in this way can personal agendas be brought into the open and addressed rather than allowed to subvert the common good. Such sharing, however, requires a trusting atmosphere, and creating and maintaining such an atmosphere is a common boss responsibility. Common bosses should advocate and promote honest expression, discourage hostile retorts and rebuttals, and ensure that messengers with bad news are not shot. Multiple bosses should support their common boss in maintaining a trusting atmosphere.

Common bosses are also responsible for the pace of planning. Appropriate planning takes time, and common bosses may have to protect their organizations against pressures from ultimate customers who seek premature product deliveries.

8. *What should the priorities be?* Whenever conflicts arise in assignments for multibossed individuals, the common boss should decide which work gets priority. The common boss should identify not only gross priorities but also priorities among the various subprojects in order to adjust resource assignments for maximum advantage. Setting priorities among subprojects depends, of course, on knowing the costs and benefits of accelerating and decelerating work that shares resources. This information has to come from the multiple bosses.

   Revising schedules will be easier whenever priorities change if multibossed individuals subdivide their work into small, pseudoindependent pieces that can be individually planned, scheduled, budgeted, and monitored.

9. *What agreements are necessary?* An agreement is needed to describe the multibossed individual's work and concomitant relationships with the multiple-boss customer and multiple-boss provider. This agreement should indicate who bears responsibility *and has authority* to assign resources, to make schedule commitments, to control expenditures, to control technical quality, etc. These variables are not independent, and they should be negotiated as a set.

   Incompatible, inconsistent, or overconstrained conditions serve no one well in the long run, and no multiple boss or multibossed individual should accept them. Since agreements are negotiated and are not unilateral dictates, they cannot be imposed against the wishes of any party. Any party can refuse unreasonable conditions and effectively stop a proposed agreement in its tracks.

Presumably, all parties want to work out reasonable agreements. The multiple-boss customer needs work done, and the multiple-boss provider and the multibossed individual want work to do in order to warrant their organizational presence. Each needs the other, and neither can ignore the other's need for reasonable conditions. This fact should act as an incentive for all parties to propose only reasonable conditions and to work toward win–win arrangements.

If multiple bosses are working within the range of reasonable conditions but are unable to agree on priorities or other issues, they should refer their differences to their common boss for resolution, as described in item 8 above. The common boss should hear the arguments on all sides and rule without undue delay. No purpose is served by stringing matters out, providing of course that the arguments do not need the support of further study. Unnecessary delays prompt multiple bosses or multibossed individuals to take matters into their own hands, typically to the detriment of the common good.

The arrangements for a given multibossed individual may involve more than one multiple-boss customer. Conflicts, if any, are likely in these cases to exist primarily between these multiple bosses and to concern the individual's availability to serve each of them. The multiple-boss providers or the multibossed individual may, however, be the first to recognize the conflict, while trying to work out the individual's assignments. In these cases, they should convene the multiple-boss customers to reconcile the conflict. The multiple-boss provider plays an essential role in this reconciliation because his or her concurrence about the multibossed individual's assignments is required for agreements with all the multiple-boss customers.

Conceivably, multiple-boss providers can establish a first-come, first-served approach to assigning their multibossed individuals, effectively saying that all assignments are intrinsically of equal merit and that priority shall be determined solely according to the order in which service is requested. While this approach is easy to administer, it is unlikely to serve the organization well overall. Existing agreements may have to be renegotiated when request sequencing alone is insufficient to determine which work shall receive the services of multibossed individuals.

Renegotiating existing agreements is not a simple matter; it may involve renegotiating commitments with ultimate customers. When this is impossible or impractical, the common boss must either arrange for more resources overall, perhaps through subcontracting, or decline new work that causes resource conflicts in the first place. A finite staff has a finite capacity for work, and extraordinary efforts cannot be sustained indefinitely (taking full account first of what is typical or ordinary effort, which may already be more than forty hours of work per person per week).

C. Matrix Behavior

A successful matrix arrangement not only involves appropriate planning and coordinating mechanisms, it also involves appropriate behaviors. This section describes these behaviors for multibossed individuals, multiple bosses, and common bosses. As mentioned previously, a project manager could have any of these roles, depending on the immediate situation.

1. Multibossed individuals
   a. Appreciate the flexibility that comes from having more than one boss. Having different bosses with different priorities and values increases the odds that one will have at least one boss sympathetic to the individual's needs, e.g., for special training or special working arrangements.
   b. Lobby to win support before decisions have to be made. Getting two or more multiple bosses to agree on an action plan that they can all support can take

more time than getting just one boss to agree. This means that multibossed individuals must allow for this extra effort and start lobbying early before decisions are needed.

c. Avoid absolutes. If a multibossed individual assumes a take-it-or-leave-it position, the odds are fairly high that at least one multiple boss will opt to leave it. Absolute positions rarely serve the multibossed individual well.
d. Negotiate matters that are in dispute. The essence of a negotiation is the exchange of concessions that leave all parties better off after the negotiation than they were beforehand. Multibossed individuals need to work toward such win–win outcomes when matters are in dispute.
e. Convene multiple boss meetings. Multiple bosses may need, but fail to establish, face-to-face meetings to resolve issues. In these cases, the multibossed individual should convene such meetings lest the issues go unattended indefinitely.

2. Multiple bosses
   a. Take a corporate point of view. Matrix arrangements are designed to optimize the overall or corporate good, not the good of any individual portion. Accordingly, multiple bosses should take a corporate, not a parochial view.
   b. Avoid being threatened by sharing subordinates, decisions, and control. Inherent in matrix arrangements is the notion that local sacrifices or compromises are made to optimize the overall good. One of the multiple bosses' sacrifices is complete dominion over subordinates, decisions, and other matters of control. Multiple bosses should not feel threatened when relinquishing this dominion; otherwise, they will behave defensively and defeat the very advantages of matrix arrangements.

c. Make the logic and importance of one's directives clear. Multibossed individuals will naturally follow clear directives over confusing or incomplete directives. If multiple bosses want to compete successfully for a fair share of multibossed individuals' time and attention, then the multiple bosses must clearly explain the logic and importance of their directives.
d. Avoid absolutes. Just as a multiple boss is likely to leave it in a take-it-or-leave-it situation, a common boss is likely to do so too if a multiple boss presents such a choice. Multiple bosses should preserve some negotiating room whenever possible in order to be able to realize the flexibility that matrix arrangements afford.

3. Common bosses
   a. "Buy" the matrix concept as the best of all possibilities for the situation. Matrix arrangements are not panaceas, but they offer the best compromise in the kinds of situations described previously. In these cases, the common boss should openly accept and endorse the matrix concept. Otherwise, the concept is likely to fail, both for lack of appropriate behavior on the boss's part and for lack of appropriate behavior on the part of others who take their cue from the boss.
   b. "Sell" the matrix concept to the members of the or ganization, are vocal and articulate in developing the concept and enthusiasm for it, and keep it sold. Matrix arrangements are not easy to maintain because they require people to behave cooperatively when some of them would rather behave selfishly or competitively. To have credibility, common bosses should acknowledge the difficulties of matrix arrangements but not dwell on them. Rather, they should promote the matrix's virtues consistently and persistently in order to establish and maintain a suitable matrix climate.

c. Maintain a balance of power among the multiple bosses. Negotiations among multiple bosses will succeed only when each party regards the other as worthy, which means that neither party can be consigned to an inferior position. Common bosses need therefore to maintain a balance of power among the multiple bosses so that none of them can treat others condescendingly.
d. Manage the decision context, see that conflicts are brought into the open and that debate is reasoned and spirited, and decide in a timely way. Multiple bosses who are left hanging by common bosses will take matters into their own hands. After all, they have jobs to do. To avoid this outcome, common bosses must decide in a timely way, using all the relevant information available. Common bosses have to manage the decision context accordingly.
e. Set standards for planning, exchanging information, and performance. Matrix arrangements work best when problems are identified and addressed early and clear standards exist for multiple bosses and multibossed individuals to follow, especially regarding cooperation. Common bosses should set and enforce appropriate standards. In particular, they should not indulge immature behavior.

8

Project Records and Reports

This chapter concerns project records and reports to the customer and to senior management. Project records facilitate decision-making and minimize the need for redoing work. They also provide the basis for reporting status and progress to the customer and to management and for estimating the effort and duration of future work. Reports are important to all customers and managers who take the project seriously. If they are not properly informed, they will not be satisfied, no matter how good the work itself may be.

I. PROJECT RECORDS

A. Organization

Project records originate throughout the project, but they should be organized and maintained as if they formed a single collection or database. This does not mean that they must be kept in a single location. As desirable as this might be, it is not always possible, especially in cases of very large projects or projects that involve different organizations that manage their own information. In these situations, the collection or database may be physically dispersed provided it is treated as a virtual single entity. Importantly, only one official copy of information should exist. All other versions are merely "working copies" that are always superseded by the official copy. Otherwise, keeping everybody synchronized soon becomes impossible.

During the project, the project's records are the responsibility of the project manager, who may of course delegate the daily work of maintaining project records to a staff member. Once the project is completed, the project's records should be turned over to the organization's archives.

B. Contents

Project records should contain at least the following information:

1. Table of contents, with a guide to their locations
2. Current version of the statement of work or product specification
3. Previous versions of the statement of work or product specification
4. Documentation of assumptions and their rationales
5. Correspondence and notes regarding agreements and understandings with functional managers
6. Correspondence and notes regarding interactions with the customer
7. Correspondence and notes regarding interactions with project staff
8. Meeting agendas and minutes
9. Action item lists
10. Current resource budgets and schedules and backup materials, e.g., estimates, calculations, quotations, etc.
11. Previous resource budgets and schedules and backup materials
12. Progress and status reports
13. Milestone review presentation materials, action item materials, and summary reports
14. Change proposals, impact appraisals, and decisions
15. Reports

16. Distribution lists for project information items
17. Miscellaneous notes
18. Cross-reference index for items not in the project records

II. PROJECT REPORTS

A. Planning for Reports

The need for various project reports and their nature, extent, and frequency should be considered when the project is planned for two reasons. First, reporting takes the time of project personnel. The efforts they spend in reporting activities must be in the project plan, schedule, and budget. Most project personnel estimate too little rather than too much time for project reporting when laying out their activities. Project managers should correct this tendency when working on overall project plans. Second, if project managers consider *all* the reporting requirements together at the beginning of their projects, they can seek ways to combine reports and thereby economize on reporting efforts and costs. While adequate reporting is essential, unnecessary reporting is wasteful.

The reporting required of a project may be specified by the contract or scope of work and by organization policy or practice. The project manager should ascertain, however, that the specifications are appropriate for the project at hand, since such specifications are often written generically, with little regard for particular conditions. Usually, this means that the project manager must ask the many-faceted customer and management about their true reporting needs and wants and then devise and negotiate a set of reporting specifications that economically serves the combined needs.

In defining true reporting needs, project managers should note the hazards of underinforming and overinforming. Underinforming leaves customers and management in the dark, wondering if scarce resources are being well spent. Overinforming floods recipients with material neither needed nor wanted, and it wastes time. Overinforming also conveys the impression that the project staff either lacks understanding of the customer's or management's viewpoint or cannot distinguish between the important and the trivial.

B. Reporting Mechanisms

Six ways a project can report its work are:

1. Formal written reports
2. Informal reports and letters
3. Presentations
4. Guided tours
5. Informal meetings
6. Conversations

1. Formal Written Reports

Formal written reports are the mechanism that most project staff, and indeed most customers, think of when they think of project reports. These, together perhaps with presentations (Section II.B.3), are likely to be specified in the contract or scope of work.

The virtues of formal written reports are:

1. They are more or less self-contained and complete and are therefore useful to personnel not familiar with the project.
2. They are in a form that facilitates their long-term retention and thereby provide a permanent record of the project.
3. They can accommodate background descriptions, literature reviews, and numerous appendices that might otherwise have no home but are important to understanding various aspects of the project.

Formal written reports also have some liabilities:

1. Because of their potentially widespread and long-term use by people not intimately familiar with the project, considerable care must be taken in their preparation. This requirement makes it difficult for them to be issued quickly or inexpensively.
2. Because the potential audience may be diverse, it may be difficult to satisfy completely the needs of all readers.

Given their virtues and liabilities, formal written reports are often reserved for reporting, analyzing, and interpreting major units of work rather than for reporting progress in real time. Often very little in a formal written report will be new information to the people who are connected with the project; these people will have already learned the essentials through other means. Nevertheless, the project manager should not shirk this activity, for it may be contractually required. Moreover, the reports may carry great weight with those who are not intimately connected with the project but rely on them as their only source of information on what was done and accomplished.

2. Informal Written Reports and Letters

Informal written reports and letters are used to keep the customer up to date regarding project progress, accomplishments, difficulties, and near-term plans. They are often prepared on a set schedule, such as daily, weekly, biweekly, monthly, or quarterly. If the regular schedule is biweekly, monthly, or quarterly, supplemental reports may also be submitted when unusually important events occur.

The informal written report or letter differs from the formal report in the audiences that it serves and its ability to stand alone. While the formal report is designed to be self-contained and meaningful even to those who have no prior contact with the project, the informal report is addressed to an audience that is assumed to be up to date as of the previous informal report. Thus, the informal report may sometimes dispense with establishing the context, presenting background material, and so on. It must, however, refresh the reader's memory, if necessary. Therefore, the longer the reporting interval, the more likely it is that some review of the project's previous status or current plan will be needed.

The virtues of informal written reports are:

1. They can be quickly prepared, so they are timely.
2. They are relatively brief, so important items are less likely to be buried and go unrecognized.
3. They can reach a large audience rather inexpensively.

Liabilities of informal written reports are:

1. Lacking full explanations and context, these reports may be misinterpreted by people who have not kept up with the project's progress.
2. As interim reports, some of the information presented may be superseded. However, the readers will not necessarily know if it has been unless they read subsequent reports.

The characteristics of informal written reports obviously make them a very useful mechanism for keeping their readership up to date. They should not, however, be considered substitutes for formal written reports.

Before leaving the subject of informal written reports, we note that some customers expect to see copies of daily field notes or laboratory notes. If this is the case, some attention should be given to how much analysis or interpretation should be sent along. One school of thought says that no analysis or interpretation should be transmitted in this way. The reason is that it may later be repudiated after additional data are obtained or further thought is given to the matter. Another school of thought holds that if the project staff do not offer an explanation, then the customer may apply its own and do so erroneously. Both arguments have their merits. Thus, rather than advocate one school of thought or the other as a general rule, we suggest that the project manager weigh the advantages and risks of each in light of the customer and the situation at hand and act accordingly.

3. Presentations

A presentation is, in effect, the oral equivalent of a formal written report. It is more than just a meeting to discuss project status, although such a discussion may follow the presentation (see Section II.B.5). The purpose of a presentation is to take advantage of face-to-face interactions (Chapter 7) while transmitting the significant aspects of a major work unit.

The virtues of a presentation are:

1. An astute presenter can adjust his or her delivery while proceeding in order to reach the audience appropriately.
2. Because a presentation is "live" and not "frozen" as is a written report, it provides an opportunity to try out alternative ways of conveying information and ideas.
3. The captive nature of an audience at a presentation may result in some people paying attention to the project who would otherwise not do so.
4. A presentation affords an opportunity for many different people to discuss a project who otherwise would not be current on its status and take the time to do so.
5. An audience may understand information to be conveyed by a graph, table, or chart better when it is presented than when it is written.

Liabilities of a presentation are:

1. The presenter must be able to retain poise when questioned by members of the audience.
2. If the presenter is not expert in all facets of the work being reported, it may be necessary to have several experts attend the presentation.

3. A presentation is not easily made to a geographically dispersed audience.
4. A presentation does not per se produce a permanent record.

It is often appropriate to use both a presentation and a written report. This approach offers the advantages of both and helps overcome some of their disadvantages.

If the presentation precedes the publication of the final report, it can be used to test ideas before they are printed or to introduce the audience gently to material they might otherwise find surprising or distressing. Also, a discussion following the oral presentation may provide suggestions that can improve the written report.

4. Guided Tours

A guided tour refers to taking customers to the project site in order to familiarize them with the work. Sometimes there is no good substitute for a first hand examination of the project situation. Yet many a customer has never visited its project site. Accordingly, such a customer has little appreciation of the project manager's situation. In these cases, the project manager will do both the project and the customer a favor by organizing a site visit.

Virtues of site visits or guided tours are:

1. They develop insight on the part of the customers which may be difficult if not impossible to develop any other way.
2. They tend to be informal, which makes it easier for the participants to get to know each other and thereby develop trust.
3. They tend to be absorbing, which means that the customers' attention will less likely be divided.
4. They tend to take a fair amount of time, which provides multiple opportunities to communicate.

Liabilities of guided tours are:

1. They are difficult to arrange if many people must be coordinated or if they are remotely located from the normal activities of the individuals.
2. They are relatively expensive to conduct, especially if distances are great.
3. They may interfere with the normal activities at the project site.

While the direct and indirect costs of a site visit are not insignificant, there are times when the visit is perhaps the only effective way to report to the customer the real essence of the project. If this is the case, the project manager should include one or more site visits in the project plan, schedule, and budget. These visits will pay the largest dividends if at least the first one occurs rather early in the project.

5. Informal Meetings

Informal meetings are often the most productive way to report to a customer. While overall they may not satisfy contractual stipulations, they are more likely to produce a two-way report, that is, from the customer as well as to the customer. As mentioned in Chapter 2, information from the customer constitutes a vital input for the project manager.

Informal meetings may either follow a presentation or be held separately. Their virtues are:

1. They are in fact informal, so that the formalities that accompany a presentation are usually missing. This enables the attendees to use their time to good advantage if they practice good meeting principles (Chapter 7).
2. The participants may be but a subset of those who might attend a presentation, so that the discussion can be limited to the needs of a limited group.
3. They generally provide a more inviting atmosphere than a presentation, so all the attendees feel able to participate, not just a few key people.

4. They can usually be arranged with relatively little fuss and only moderate advanced notice.

The main liability of informal meetings is that all the right people might not be present, so some decisions may have to be reviewed or reconsidered before they can be implemented.

Needless to say, the principles of good meetings discussed in Chapter 7 for project staff meetings apply to meetings between the project and its customer. There may be a slight problem, however, in deciding who is in charge of the meetings.

When informal meetings between the project and its customer are called at the request of the project manager, the customer's chief delegate may nevertheless want to be in charge. If the latter chairs the discussion, the project manager may or may not get the issues resolved satisfactorily, depending on the skill of the chair. If the project manager suspects that the chair will not be effective, he or she may try to arrange in advance to lead the discussion once the introductory remarks have been made. The chair can then summarize and conclude the meeting after the discussion is over.

6. Conversations

Conversations have the same advantages in reporting to the customer that they have within the project organization (Chapter 7). They are, however, hardly ever recognized as an official vehicle for the project to report to its customer. They are simply too casual to satisfy interorganizational responsibilities. This does not mean that they cannot or should not be used.

Rather, conversations should be used between the project and its customer to establish understandings. Once these understandings have been reached, they should be confirmed by an informal report or a letter (see Section II.B.2).

In this way, the project manager can have the benefits of conversations and still practice good reporting technique.

C. Choosing a Suitable Reporting Approach

In choosing a reporting approach, the project manager should consider the nature of the audience and their needs and the time and resources available. The audience can be characterized along five dimensions:

1. *Audience diversity.* It is difficult to serve a widely diverse audience with a presentation. One part of the group may be bored while the needs of another part are being addressed. The other approaches listed in Section II.B can all accommodate a diverse audience if suitable care is exercised.
2. *Audience sophistication.* An unsophisticated audience may benefit most from an informal approach—whether in writing or orally—that borders on the tutorial. And an oral approach, with two-way interaction, is generally more effective in this case than an informal written report. A more sophisticated audience may prefer a more formal report or presentation which is then followed by discussion.
3. *Audience familiarity.* If the audience is familiar with not only the project as described in the scope of work, etc., but also the project site, then written and oral reports and presentations will probably be effective. On the other hand, if the audience is not familiar with the project site, then a guided tour is in order.
4. *Audience size and geographic location.* If the audience is large or widely dispersed, then written reports may be the only viable means. But if it is compact in both size and location, then presentations and informal meetings should be considered for the two-way benefits that they offer.
5. *Audience need to know.* Do the members of the audience all have the same need to know? If so, presentations and informal meetings may be effective. If not, written re-

ports are probably better, providing they make clear where each segment of the audience will find what it needs.

Project managers will probably feel constrained by the schedule, budget, or availability of key personnel as they try to optimize their reports to clients. They need, therefore, to consider ways of compromising that will not at the same time seriously harm their projects.

They should consolidate reporting efforts where practical in order to minimize redundant work. They should schedule reporting activities at opportune times, not necessarily according to rigid schedules, so that project staff can both perform their work and report on it with a minimum of extraordinary stress. And they should arrange reporting events, especially presentations, well enough in advance that the reports are timely when received. This latter practice will minimize the need for ad hoc reports to selected individuals who would otherwise request or require special reports.

D. Preparing Reports and Making Presentations

This book is neither a text on expository writing nor a treatise on public speaking. Yet it seems in order to present a few tips on effective reports and presentations that are frequently overlooked in the heat of project activity. These few suggestions together with some basic communication skills will help project managers report successfully to their customers.

1. Plan the report or presentation. What are the objectives for the communication? What does the audience already know? How can one get the audience from their present levels of understanding to the objectives? First, the route should be outlined. Then, the details should be filled in.
2. Capture the audience's interest at the outset, then develop the subject. If the subject is developed before their

interest is captured, they may not pay sufficient attention to get the message.

3. If preparing a written report, be sure that all figures can be interpreted without reference to the text. Also, figures should not require color reproduction to be interpreted, since many figures are later photocopied in black and white.
4. If making a presentation, be sure that the physical conditions are suitable. Otherwise, they will distract the audience. The air temperature and quality, noise and lighting levels, the quality of seating, and the audience's ability to see any audio-visual aids and to hear the speaker all deserve attention.
5. Practice good presentation techniques:
   a. Define and post the objectives for the presentation and tell the audience why the objectives are important. Do not assume that the audience knows either the purpose or why it is significant.
   b. Tell the audience what the path or topic sequence will be. They will receive the elements of the delivery serially. Since they cannot ruffle through the delivery like a book to see where various elements come, they need to be told in advance in order to put the pieces in place as the presentation proceeds.
   c. Use visual aids as appropriate. They should help the audience follow and understand the presentation but they should not be a substitute for it. That is, the audience should get more from the total presentation than they get from the visual aids alone. See also item 6.
   d. Build bridges from one topic to the next to help the audience follow the presentation.
   e. Do not give needless details. Many members of an audience are offended when they receive more information than they need.

f. Sum up major points so that the audience will recognize them.
g. Recommend action when it is appropriate. Most audiences need a concrete proposal to focus on when they contemplate future action.
h. Watch the audience's reaction to the presentation and adjust it to get desired results. Avoid a stilted delivery that puts the audience to sleep. Establish and keep credibility by being candid and forthright. (And keep enough lights on in the room to be able to read the expressions and other nonvocal cues of the audience.)

6. Use visual aids appropriately:
   a. Use lettering and symbols large enough and in sufficient contrast with the background that they can be read from the back of the room. Demonstrate that they are satisfactory *before* the presentation and make corrections as necessary.
   b. Put related items on the same chart if doing so does not result in crowding the chart. Do not distract the audience with unrelated material. Sometimes cover a list on a chart and expose each new item only when it is time to discuss it. This technique is helpful when the list is long and the time devoted to each item is also significant. The technique keeps the audience from reading ahead to later items while an earlier item is still being discussed.
   c. Do not merely read a chart to the audience. Explain or amplify it. Give examples. Otherwise, the presentation can be reduced to a silent showing of charts. At the same time, use the same kind of language as the charts. Do not force the audience to translate one or the other to see their equivalence. The object is to have the visual aid reinforce the message and increase audience understanding, not increase confusion. Provide copies of the charts for the audience to use in note-taking.

E. Obtaining Feedback

It pays project managers to obtain feedback from those who receive their reports and presentations. Feedback enables project managers to determine if their materials were understood and provides an opportunity to correct misunderstandings if they are detected. Feedback often includes information that is helpful for future action and enhances project managers' abilities to satisfy their customers. And requesting feedback helps project managers show that they care whether or not their audiences understand.

Feedback can be obtained by interviewing, by questionnaire, or both. If interviews are used, the project manager can talk with members of the audience "cold" or can tell them that he or she will phone or visit at a particular time to discuss their reaction to the report. The latter may also defuse any tendency that members of the audience may have to send an angry letter.

A questionnaire enables the project manager to reach a larger audience, but it requires very careful preparation if it is to elicit useful information and also not offend the respondents. In fact, most questionnaires need to be "dry run," perhaps several times, to make them unambiguous, efficient, and helpful. They are generally used therefore when so many people must be surveyed that the cost of preparing a suitable questionnaire is less than the cost that would be incurred if the same people were to be interviewed.

When a questionnaire is in order, it may also be supplemented by interviews of selected individuals to learn about their special reactions and needs. These individuals may be people who have a special relationship to the project or people who raise special points in their replies to the questionnaire. The latter, of course, may be detected only if the questionnaire solicits comments in addition to asking the respondents to choose from predetermined answers.

In addition to obtaining feedback from the audience of a report or presentation, the project manager can also query the project staff. They, too, need to be shown that the project manager cares what they think, and they may have useful information that they would not volunteer without being asked.

9

Negotiating

To negotiate is to confer with another to settle some matter. Project managers frequently must negotiate with clients, bosses, functional group managers, vendors and subcontractors, support group managers, and their own project staff in order to resolve issues such as scope of work, standard of work performance, staff assignments, schedule, budget, and so forth. Negotiating is so pervasive in project work that the project manager needs to know how to negotiate successfully. This chapter, therefore, describes the elements of negotiating, how to prepare for a negotiation, arrangements for negotiating, and negotiating techniques.

I. ELEMENTS OF NEGOTIATING

Certain elements are common to every successful negotiation: cooperation, something for everyone, satisfaction of real needs, common interests, and a reliable behavioral process.

A. Cooperation

Negotiating is a cooperative enterprise in which two people or parties search for an arrangement that leaves both of them better off than they were when they started. Essentially, each party makes concessions in such a way that the concessions that it receives are worth more *to it* than the concessions it gives up. The aim is to find a win–win arrangement and to avoid win–lose arrangements. Searching for a win–win arrangement indeed requires cooperation, and negotiating is thus a cooperative, not a competitive process.

B. Something for Everyone

Each party in a successful negotiation has to let the other party win something of value to it. Otherwise, that party has no reason to participate and may withdraw from the negotiation. If one party withdraws, then neither party can win anything. That is, concessions cannot be obtained from a party who refuses to negotiate.

If one party enters a negotiation with the idea that it will utterly vanquish the other party, it is not psychologically prepared to negotiate. As soon as the other party discovers that it cannot win something of value, it can and should terminate the negotiation. Neither party should continue to negotiate (i.e., continue to trade concessions) if it is receiving less than it is giving. Thus, each party must be prepared to let the other win something.

C. Satisfaction of Real Needs

A successful negotiation satisfies the real needs of the two parties. A real need is one that is considered genuine by the party who is presumed to have the need, not the party who is offering the concession.

To put the matter another way, there are rewards and nonrewards. A nonreward is not a punishment; it is just not a reward. Something which the other party does not value is a nonreward, no matter how valuable or rewarding the offering party thinks it should be. Nonrewards do not satisfy needs, whereas rewards do. Each party must search for rewards that the other party will value if it hopes to negotiate successfully.

While a party can offer a nonreward, it will not advance the negotiation nor induce any concessions from the other party. It may, however, be accepted by the party to which it is offered. If so, the offering party gives up something needlessly and gets nothing in return. Overall, it may be worse off for having offered the nonreward.

A key skill of a successful negotiator is the ability to determine quickly and with some accuracy the needs of the other party. One thus avoids offering schedule relief when the other party truly needs more budget. One does not offer extra financial gain when prestige is the real issue. And one does not offer security when self-actualization is the other's objective. By the same token, one does offer schedule relief when possible *if* it meets the other party's need. And so forth.

In assessing another's needs, it is risky to impose one's own standards. We have all heard of artists who live happily in hovels while intently pursuing their art for little or no pay. We might believe that their living conditions indicate a need for better housing. This is not necessarily true, and an offer of better housing could be a nonreward for them. Simple, even primitive, shelter can satisfy their needs while it might not satisfy ours.

The other party's needs and constraints can be inferred from a variety of sources:

1. Budgets and financial plans
2. Publications and reports
3. Press releases
4. Officers' speeches and public statements
5. Institutional advertising

6. Reports to agencies, e.g., the Securities and Exchange Commission
7. Moody's and Standard and Poor's reports
8. Recorded real estate deeds
9. Credit reports
10. Instructional and educational materials
11. The other party's library
12. Biographical references, e.g., *Who's Who*
13. Others who have faced the same party

The sources may suggest rather concrete needs, such as the need to complete a project by a particular date or the need to accomplish certain work at the lowest possible cost. But some sources may also suggest less tangible objectives, such as being master of one's own affairs, being regarded as an influential person, or being protected from excessive risk.

D. Common Interests

Negotiating involves seeking common interests. The two parties need to identify together what each wants and what each can give up. It is best if such information is traded in small amounts; this approach helps keep a cooperative spirit going until agreement is reached. Also, trading information in small amounts helps each party avoid offering a substantial concession without receiving a commensurate concession in return.

E. A Behavioral Process

Negotiating involves trading information and concessions in a somewhat formal way. It is a little like a minuet, with each party behaving in a way that is understood by the other. This behavior enables both parties to estimate about how much information or concession to offer at any moment and how much to expect in return. Each party thus feels that the concessions that they are exchanging approximately balance. Each is therefore willing to continue the negotiation until both are satisfied with their overall situations.

II. PREPARING FOR A NEGOTIATION

A project manager can generally enhance the chance of a successful negotiation by preparing for the event. This section describes several steps that can improve the likelihood of a successful outcome.

A. Determine Objectives

The project manager should determine the objectives for the negotiation and should know which aspects of the desired results are absolutely essential and which are merely nice to have.

Some assessment should also be made about the difficulties to be expected in trying to obtain the various objectives. This assessment will be useful in establishing an agenda for the negotiation (see Section II.D).

B. Determine What Can Be Yielded

Of those items that are wanted, the project manager should determine which can be given up if doing so would meet a need of the other party. They should be ranked in priority so that the less "expensive" ones can be offered before the more "expensive" ones (as seen by the offeror). This will enable the negotiator to offer items of small value to it in order to induce concessions and build goodwill in the early stages of the negotiation (see Section IV).

Also, each party should determine what cannot be yielded, to avoid inadvertently yielding an element that it considers essential.

C. Determine the Other Party's Real Needs and Constraints

Each party should understand the other party's real needs so that it can try to meet them. It should also understand the other's real constraints, so that it does not make demands that cannot possibly be met. If the other party is aware that it is being asked to violate its own real constraints, it may withdraw from the negotiation, for it is bound to lose if it stays in.

D. Prepare a Tentative Agenda for the Negotiation

The project manager should prepare a tentative agenda for the negotiation. Since the other party should also prepare a tentative agenda, the first item on the agenda, after introductions and preliminary remarks, should be to agree on the actual agenda that will be followed.

Generally speaking, the agenda should progress from minor items and items that will be easy to agree upon to major items and items that may be difficult to agree upon. The point here is that each party needs to develop a feeling of cooperation and goodwill toward the other before attempting to negotiate major and difficult issues. This is most readily done when small concessions are exchanged in the early stages. Then, by the time the difficult issues are reached there will be a history and a spirit of satisfying mutual interests. This experience and attitude will help the two parties accommodate each other where possible in order to reach a mutually satisfactory conclusion.

E. Determine Your Image in the Other Party's Eyes

A party to a negotiation should try to determine how it is seen by the other party. Suitable forewarning can help it prepare to combat unproductive behavior by the other side.

For example, if the other party thinks that it is facing a pushover, it may make extraordinary demands. Then, it may feel honor bound to obtain these demands even if it later learns that they cannot be granted. The situation becomes a matter of saving face for the party making the demands and it can stymie the negotiation. This type of impasse can be removed by offering an alternative approach instead of just denying the requested concession. Or, occasionally the request can be broken into several parts and some of them given up.

If the other party thinks that it will be dealing with a very tough negotiator, it may behave defensively and be reluctant to exchange information that is vital to concluding a successful negotiation. In this case, it pays to treat the other party warmly and to build its confidence in one's own reasonableness.

If one expects the other party to err in assessing its opposition, then it is well to try to revise the other's impression from the beginning. This can be done when the negotiation is being arranged and when the agenda is being determined. One can, for example, show that he or she is reasonable to deal with by accommodating the other party on issues that have little significance. At the same time, one can display the ability to stand steadfast to principles when there are, in fact, important positions to protect.

F. Develop Alternative Solutions

A party about to enter a negotiation should develop alternative acceptable solutions to the difficult issues in case its first approach is unsatisfactory to the other party. Since common interests are to be sought, some exploration of alternatives is in order during the negotiation. The proceedings can be facilitated if each party knows in advance some alternatives that are acceptable to it and those that are not. Such preparation will enable it to suggest or focus attention on acceptable solutions and to steer away from unacceptable solutions.

G. Decide Who Should Negotiate and Assign Roles

Selecting a negotiator for a particular negotiation involves two decisions. One decision is whether the negotiator should be an individual or a team, and the other is who it or they should be. If a team is chosen, role assignments for the team members must also be decided.

A single negotiator has certain advantages and disadvantages:

1. It avoids questions being asked of weaker members, since there are none, and thereby avoids having their answers undermine the work of the team leader.
2. It avoids disagreement among team members which the other party could use to attack the leader's arguments.
3. It facilitates on-the-spot decisions to make or gain necessary concessions.
4. It places complete responsibility in one person, except as he or she is restricted in authority (and therefore required to seek the approval of others who are not present at the negotiation).

Team negotiation also has some advantages and disadvantages:

1. It can provide people with complementary expertise to correct misstatements of fact.
2. It provides witnesses as to what was said.
3. It enhances the reading of the other party's body language by having more observers.
4. It enables the pooling of judgments.
5. It presents the other side with a larger opposition.
6. It requires one person to be designated the chief negotiator, who serves as the focal point for the team, directs the participation of individual team members, and makes on-the-spot decisions if necessary.

Generally speaking, if the members of a negotiating team are well instructed and well disciplined, the team approach is the better of the two when major issues are involved or the outcome is crucial. Even a two-person team can be much more effective than a single negotiator. A team approach assumes, of course, that the team members are available for the duration of the negotiation.

A single negotiator may be better when the issues are relatively minor; when they must be resolved promptly, that is, without making formal arrangements involving several people; or when a "public" exercise might cause one party or the other to behave unproductively in order to be consistent with previous positions or otherwise save face.

An important factor in selecting a negotiator or negotiating team is the relative position of the other party's negotiator(s). Contrary to a commonly held view, one should not send a negotiator or team that has more authority than the other party's negotiator(s). Since negotiation is largely a matter of exchanging concessions, one does not want to be in the position of being able to make larger concessions than the other party can give in return. The argument applies to the other party as well. Thus, in a well-planned negotiation, the two parties will send negotiators whose authorities are approximately the same in terms of the kinds of commitments that they can make.

In the event that one does send a negotiator whose authority is significantly greater than that of the other party's negotiator(s), the negotiator with "excess" authority can simply indicate that he or she must have the subsequent approval of others (who might be named) before proposals, and so forth, become commitments.

The need to designate a chief negotiator in a team negotiation has been mentioned. When a team is used, other roles should also be assigned. If issues of different types are involved, (e.g., technical, financial, liability, schedule, etc.), then someone should be designated to negotiate each type. This does not mean that one person cannot or should not do them all. It merely means that the entire team should know who is going to do each type.

Those who are not carrying an active role at the moment should observe the other party's negotiators. The aim is to determine from spoken remarks and from body language who is responding favorably, neutrally, or unfavorably to the concessions and commitments being proposed. This information can then be exchanged among the team members during caucuses in order to devise successful strategies. Each person of the other party should have a designated observer from one's own team. Some team members may of course have to observe more than one person.

A team member should also be designated as note-taker for the team in order to capture details that may otherwise escape recall. The note-taker needs to understand the essence and nuances of the negotiation in order to pick up key points. It also helps if he or she can glean essential information from simultaneous conversations in the event more than one person talks at once. Observational notes and other ideas should also be jotted down by the other team members as they occur so that they can be referred to and discussed in caucuses.

Finally, assignment of roles includes instructing the team members on how to behave when they are and are not the lead negotiator at the moment. The lead negotiator needs not only to conduct the discussion at the time, but also to be on the lookout for signs from team members that they need to communicate with the leader before proceeding. The leader should not go so fast so as to preclude receiving their inputs or force them to interrupt in order to steer the discussion away from unattractive areas.

By the same token, the other team members need to indicate when they have an input to make, first to the team leader and then perhaps to the entire group. They might do this by a signal of the hand or eye, or they might pass a note. Sometimes they might ask for a moment to confer privately with the leader. Generally, they should try to get the leader's attention without interrupting. There are times, however, when the leader seems oblivious to the more subtle approaches and an interruption is necessary.

A team member should not risk speaking directly to the other party without first getting an okay from the leader at the moment unless there has been prior agreement that he or she is the spokesperson on the subject or aspect at hand. If there is any uncertainty about authority on the matter, the team member should seek the leader's approval before speaking.

Team negotiations require exceptional teamwork. As in rowing, no one is praised for rugged individuality.

H. Practice

Practice is recommended before a serious negotiation for two reasons:

1. It helps identify the difficulties that will be faced in the negotiation so that alternative approaches can be developed and evaluated.
2. It permits the would-be negotiators to rehearse their roles and develop some skill in exercising them.

Obviously, the more skillful and knowledgeable the negotiators, the less practice they need. But even the most experienced negotiators generally profit from at least one dry run. This allows them to heighten their awareness of the issues and, if it is a team negotiation, to improve their coordination and ability to work together.

An effective way to practice is to have personnel who will not be part of the negotiating effort roleplay the other party in a mock negotiation. If an experienced negotiator can be enlisted as a critic and coach during the mock negotiation, so much the better.

In general, the amount of preparation needed for a negotiation depends upon several factors:

1. The skill of the negotiator(s)
2. The relative importance of the outcome
3. The alternatives available

Many negotiations are conducted on the spur of the moment, with virtually no obvious preparation. However, chance favors the prepared mind. If project managers are aware of and pay attention to the issues discussed above as they proceed, they will enhance their results compared to spur-of-the-moment negotiations. For more formal arrangements, they can and should take the trouble to prepare more formally.

III. ARRANGEMENTS FOR NEGOTIATING

The arrangements for a negotiation should be chosen with the idea that they can help attain one's negotiating objectives. The variables are the time, the place, and the schedule for the negotiation.

A. Time

When negotiating should begin is an important factor. Adequate opportunity should be allowed for both parties to prepare for the negotiation. Attention should be given to when the parties are available to negotiate. And the negotiation should be conducted and concluded in time for the outcome to be useful to the project.

In many cases, preparation, availability, and time constraints give conflicting answers about when negotiating should begin. Also, the two parties may have different needs, so they may have conflicting ideas about when to begin. When either or both of these conflicts exists, a compromise is in order.

A compromise on the starting time should not be accepted, however, if it would disadvantage the accepting party to the point that it would be better off if it did not negotiate at all. It requires courage to decline to negotiate because the terms under which it would have to be done are unacceptable. But discretion is often the better part of valor, and that is true here.

B. Place

The selection of the place for a negotiation involves such factors as freedom from interruptions; psychological advantage; and access to personnel, equipment, files, and services. The first factor is self-explanatory, but the latter two warrant some discussion.

Some psychological advantage generally accrues to the party that negotiates on its own premises. For this reason, most negotiators like to work at their own place of work or business. Since this is true of both parties, they may choose to favor neither party and instead hold the negotiation at a neutral site.

There are times when the negotiator should offer to negotiate at the other party's place of work or business. For example, if the other party would be less defensive and more cooperative if it felt more at home, then it might pay to offer to negotiate at its place. And if it would serve to build goodwill by meeting at the other party's place, then the negotiator may offer to do so. In either case, however, a negotiator should not make such an offer if doing so would create a serious disadvantage.

Sometimes it is helpful if a negotiator can control access to other personnel and files during the negotiation. One may not want to have certain experts or information readily available if finessing the issues would give an advantage. There are times when the other party will not pursue an unattractive element if it is not available or will give the benefit of the doubt if the facts cannot be easily obtained.

For this reason, a negotiator may prefer not to negotiate on his or her own premises.

On the other hand, a negotiator may choose to negotiate where selected personnel, equipment, or files are available because they are needed to make certain arguments and they cannot easily be taken to another site. Likewise, a negotiator may prefer to negotiate on his or her own premises because it has conference telephone call facilities, stenographic services, or even just a pleasant atmosphere that may enhance or expedite the proceedings.

C. Schedule

Schedule refers to the pacing of the negotiation. Generally, whenever a party feels some urgency about concluding the negotiation, it is likely to make concessions more readily than it otherwise would. It pays therefore not to be the party with the earlier deadline for concluding the negotiation.

Sometimes a party would like very much to conclude a negotiation by a particular date or time. However, it should not reveal this fact. While it can indicate its hopes if asked, it should make very clear that it is able to take as long as necessary to work out a mutually satisfactory agreement. Otherwise, the other, more relaxed party may stall until the deadline approaches. As the deadline nears, the more anxious party may lose patience and become unwilling to work for what it wants. It can end up giving concessions more easily and demanding less in return than it would if it were not so anxious.

Even when one party does not press the other against a deadline, the negotiation may need to be concluded promptly for the result to be timely. When this happens, one party (or both) may be unable to strike the bargain that it would like to have. It should not be disappointed, however. When half a loaf is better than none, the correct comparison is between something and nothing, not between something and everything.

By the same token, a party should not let schedule deadlines force it to negotiate a truly unsatisfactory result. It should instead withdraw from the negotiation. As mentioned in Section I.B, if one cannot be better off after the negotiation than before, there is no reason to participate.

IV. NEGOTIATING TECHNIQUES

The term *negotiating techniques* has two meanings. The first meaning refers to practices or techniques that help negotiators change information and concessions and help them avoid making unwanted concessions. The second meaning refers to psychological pressures that one party may apply to the other in order to obtain concessions that the latter would not make in the absence of the pressures. Techniques of the first type are discussed in this section. Information on techniques of the second type can be found in the popular literature.

A. Defining the Issues

It is important for the negotiators to define the issues to be negotiated. Each party should name its concerns and propose a sequence or an agenda for considering the issues. The two parties should agree upon a common agenda before proceeding to the issues themselves.

As discussed in Section II.D and Section IV.C, it is generally helpful if the issues that appear difficult to either party are placed toward the end of the agenda. Thus, each party should let the other defer whatever issues it wants. If this is done, only issues that seem to both parties to be minor or easily resolved will come first. As the easy issues are resolved, both sides will build a reservoir of goodwill toward each other that will help them resolve the more difficult issues.

If one party wants to negotiate an issue that the other feels is non-negotiable, it is best to put the issue on the "difficult list" and defer its consideration until after the other issues have been resolved. Perhaps the second party will change its mind if everything else goes well. Even if it does not change its mind, it can wait until the issue is about to be actively discussed to declare that it is non-negotiable.

B. Establishing Participant Authority

Since negotiating consists essentially of exchanging concessions and each party wants to receive at least as much as it gives, it is important for the two parties to have commensurate authority to make binding concessions. Neither party should be empowered to give greater concessions than it can receive from the other. Otherwise, the party with the greater authority will find itself making significant concessions that it must honor while the other party will have to honor only less significant concessions.

To avoid making more significant concessions than it can receive, each party must know the authority of the other. If one party determines that it has more authority than the other, it can shed some of its authority for the time being by declaring a need to consult with absent personnel (see Section IV.D).

If one party does not know the other's authority to make binding concessions, it should ask. Simple, straightforward questions are best, such as, "To what level can you make a commitment without the approval of others?" or, "Are there any areas that we are about to discuss that must be ratified before our agreements are binding?"

C. Building Goodwill

Goodwill facilitates the negotiating process. At some point, one of the parties has to offer a concession in order to induce the other to give one in return. There is no assurance, however, that the offer will not be rejected. The very prospect of rejection can inhibit the first party from making the offer. This inhibition is diminished markedly, though, when there is a feeling of goodwill that keeps a potential rejection from being taken personally.

Goodwill is developed by the parties behaving in a caring way toward each other. This can occur naturally as each party accommodates the other on details of the arrangements that are not of great importance to it.

Goodwill is also developed by resolving some issues easily. When agreement is reached, the two parties usually feel good about their accord, for it reduces some of the tension and anxiety that accompanies all negotiations.

As a reservoir of goodwill is developed, it is easier for the parties to cope with the tensions that accompany the more difficult issues. Indeed, when the goodwill is adequate, it may be as easy to deal with a difficult issue as it was to deal with an easy issue before there was significant goodwill.

D. Seeking Approval of Absent Personnel

The approval of absent personnel is sought when a negotiating party either truly lacks authority or wishes to avoid having authority to make binding commitments on the spot (see Section IV.B). All commitments that are made subject to another's approval are, of course, subject to being reversed.

If one finds that the other party needs to have its tentative "commitments" confirmed by an absent person before they become binding, then one should make one's own "commitments" conditional upon the confirmation. If the confirmation is not forthcoming, then one is relieved of the conditional "commitments."

E. Seeking Higher Authority

Occasionally one party finds the other intractable and wants to deal with a higher authority. If so, the first party should ask to talk with the other party's superior. The request is likely to be granted.

If the other party asks to talk with one's own higher authority, the higher authority should be briefed thoroughly before entering the negotiation. Moreover, he or she should not be allowed to negotiate alone but should be accompanied at all times by someone who has attended the negotiations up to that point. This will avoid the other party's trying to renegotiate old issues.

F. Interrupting Unattractive Lines of Pursuit

Sometimes a party wants to avoid discussing a presently favorable but indefensible position. When this happens, the first party wants to interrupt the other party's line of pursuit and do so with a minimum of fanfare. This will minimize the chance that the issue will be raised again.

Possible means of interrupting are to call for a coffee break, ask for a caucus, adjourn for lunch, distract the conversation, go to the rest room, and so on. Upon resuming the meeting, one can begin with another subject, thereby making it difficult to reintroduce the troublesome topic.

When there is a team negotiation, the chief negotiator should warn colleagues about the possibility of having to interrupt unattractive lines of pursuit. Otherwise, one of them may unwittingly continue or reintroduce an unwanted issue in an attempt to tie up loose ends.

G. Securing Commitments

Both parties to a negotiation need to confirm in writing what they have agreed to. This agreement should be done on the spot at least session by session, if not issue by issue. These written agreements prevent reneging on and arguments about the results of earlier discussions. They thus facilitate forward progress of the negotiation.

If no other means is conveniently available, the agreements can be captured on a flip chart or just written longhand on paper. Both parties can signify their agreements with the written versions by initialing them.

Epilogue

Project management is an integrating activity and is done best when the manager has a holistic outlook. A typical project involves numerous conflicting pressures and requires artful compromises. Successful compromises can be made only when the project manager understands how all the elements of the project interact and work together to accomplish the project's objectives.

For ease of explanation and understanding, many of the chapters in this book deal with individual aspects of project management. This is not to imply, however, that these aspects stand alone or that they can be pursued independently. For example, one usually cannot plan without some negotiation, and one cannot negotiate without some prior planning. One cannot control without planning, monitoring, and directing and coordinating. Nor can one report without monitoring. Thus, we urge our readers to blend the several ingredients thoroughly in applying them to their projects.

Many projects get in trouble, and some become utter failures, that is, time and money are spent without achieving the objectives. While the reasons claimed for trouble or failure are varied, they generally boil down to inadequate project management.

Sometimes project managers accept assignments under impossible conditions, perhaps by overselling themselves or buying bills of goods. Competent project managers will not deliberately work with such odds, and they structure their objectives and preliminary studies to protect against ultimate failure.

Sometimes project managers fail to exercise their prerogatives. They may, for instance, accept changes in scope without getting prior approval from their customers for the consequential changes in budget or schedule; later they are late and overbudget and may be unable to collect for the extra efforts. Or they may not ask for timely, meaningful reports from their task leaders; then they are unable to coordinate their work. Or they may hide their problems from management, who are therefore unaware of their special need for resources.

Sometimes project managers dupe themselves into thinking that they are charismatic or omnipotent and able to surmount any shortcoming, whether poor estimates, sloppy plans, inadequate contingency allowances, no controls, schedule compression due to late starts, or ineffective communications. Only toward the end of their projects do they realize that their powers are finite and quite likely insufficient to rectify the shortcomings. It is perhaps sad to see project managers lose their naivete and discover that they are not omnipotent. But the true tragedy is the waste of their project teams' efforts and their customers' resources. They are people of good intent and goodwill, and they have labored in vain.

Project managers are in a position of trust. They are not only the focal points for their projects, they are also the stewards of all the talent and resources committed to their projects. They have a noble responsibility.

Successful project management is a work of art and a joy to behold. A project failure can cause much tragedy. We hope that this book will help you manage your projects successfully and give you, your project teams, and your customers much happiness and satisfaction.

Appendix A

Work Breakdown Structure Paradigms and Processes

Project work can be represented by work breakdown structures (WBS) such as those shown in Figures A.1 and A.2. These figures are hierarchical *composed of* charts, where the items at one level of the chart are needed either directly or indirectly to make the item at the next higher level. Direct relationships are shown by a strict vertical structure, as in Figure A.1. Indirect relationships are indicated by short dashed vertical lines, as in Figure A.2, which signify that some outputs from the left of the dashed line are needed as inputs to make some items on the right. This latter relationship means that the items on the right cannot be settled until the items on the left are settled.

Figures A.3a through A.3g are WBSs for a project whose overall product is a system. The WBS in Figure A.3a shows the first level of decomposition, and the WBSs in Figures A.3b through A.3g show decompositions of the plan, the requirements, the design, a typical subsystem, the integrated system, and the validated system, respectively. Each of these latter WBSs is discussed in turn.

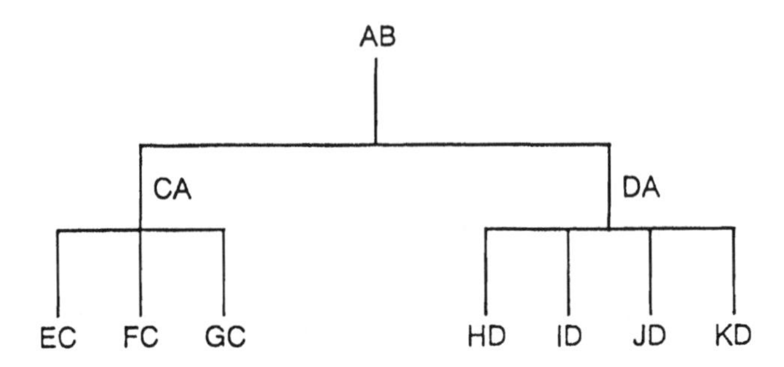

Figure A.1. A simple work breakdown structure.

1. The Plan

Figure A.3b is a WBS of the plan. This figure shows not only the components per se of the plan but also the staging needed to develop the plan, using short dashed vertical lines to separate the stages. Using a convention of time increasing from left to right, the chronologic relationships at the first level of decomposition are as follows:

a. *Initial objectives and constraints* is the left-most element at the first level of decomposition and is the first element at this level to be prepared.

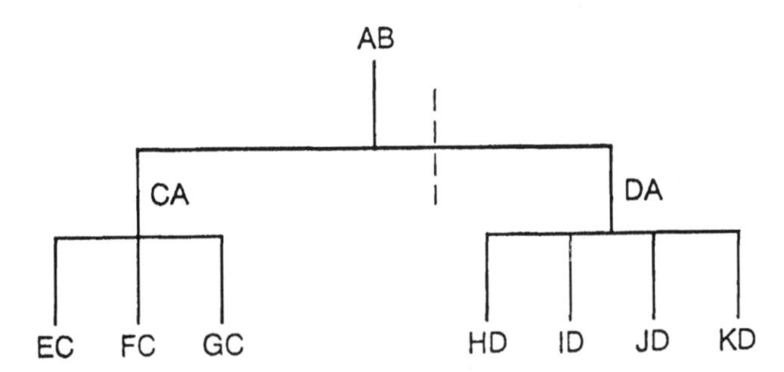

Figure A.2. A work breakdown structure with indirect relationships.

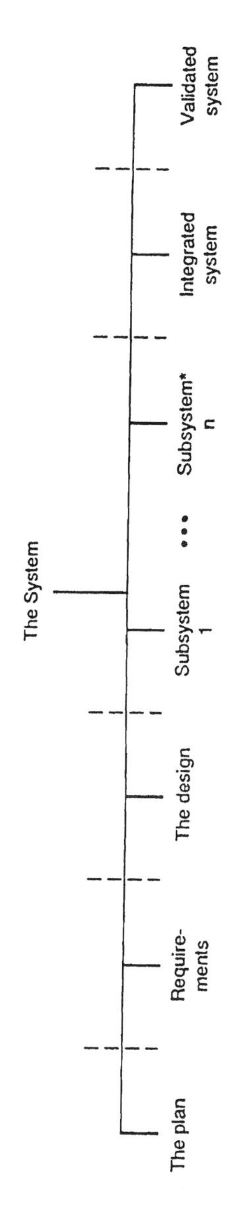

* Subsystems may include system assembly or integration tooling, test equipment, etc.

Figure A.3a. First-level decomposition of a system development project's work breakdown structure.

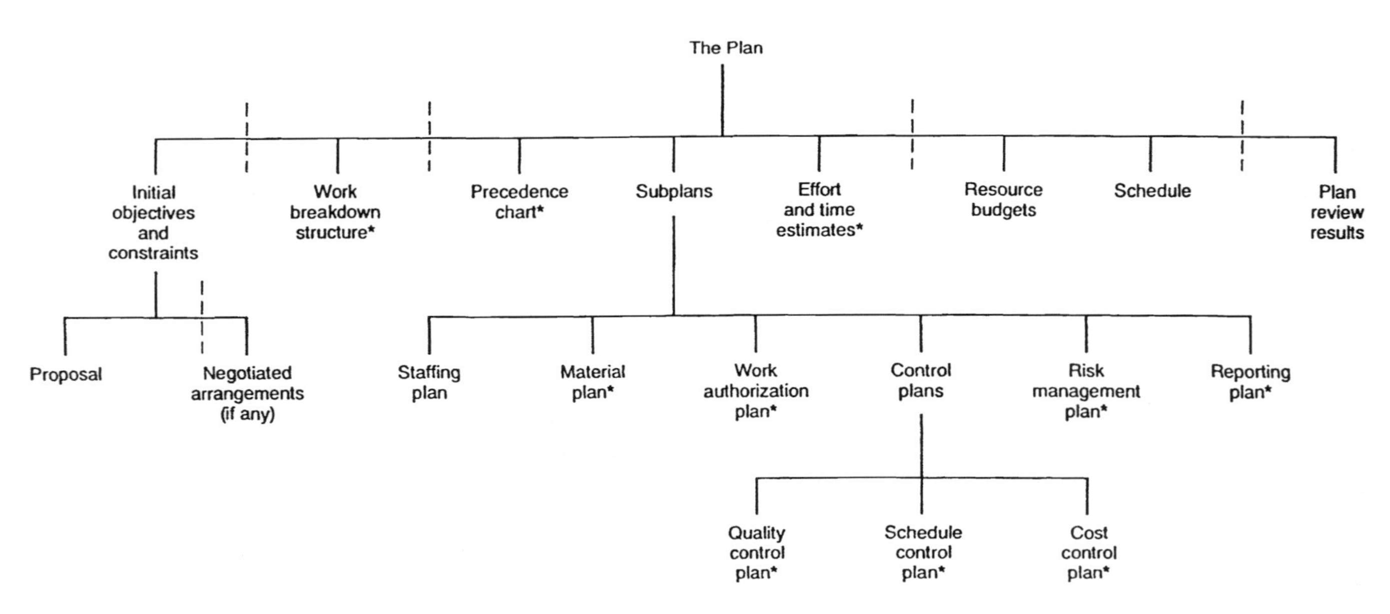

Figure A.3b. The project plan portion of a system development project's work breakdown structure.

b. *Work breakdown structure* is the next element (from the left) at the first level of decomposition and is the second element at this level to be prepared.
c. *Precedence chart, subplans,* and *effort and time estimates,* have no dashed lines between them and comprise a set of elements that are developed in parallel in the third stage of preparing the plan.
d. *Resource budgets* and *schedule* also have no dashed lines between them and comprise a pair of elements that are developed in parallel in the fourth stage of preparing the plan.
e. The six parts of *subplans* are developed in parallel, and the three parts of *control plans* are developed in parallel. In the final analysis, this means that *precedence chart,* the three individual control plans, the five non-control subplans, and *effort and time estimates* are all developed in parallel. At the same time, the structure in Figure A.3b puts the control plans together and so that they can be coordinated as a set before they are coordinated with other subplans.
f. If any later development turns out to be incompatible with the output of an earlier development, then both earlier and later parts are iterated until compatibility is achieved.
g. The plan is ready for review when all its own elements are mutually compatible. A successful review (see Appendix E) means that the plan is satisfactory, at least for the time being. As the rest of the project unfolds, the plan may have to be revised to keep the entire project in harmony.

2. Requirements

Figure A.3c is a WBS of the requirements. This figure shows the requirements not only in their final state, i.e., *requirements by type,* but also in their intermediate state, i.e., *requirements by source.* The figure also shows *requirements re-*

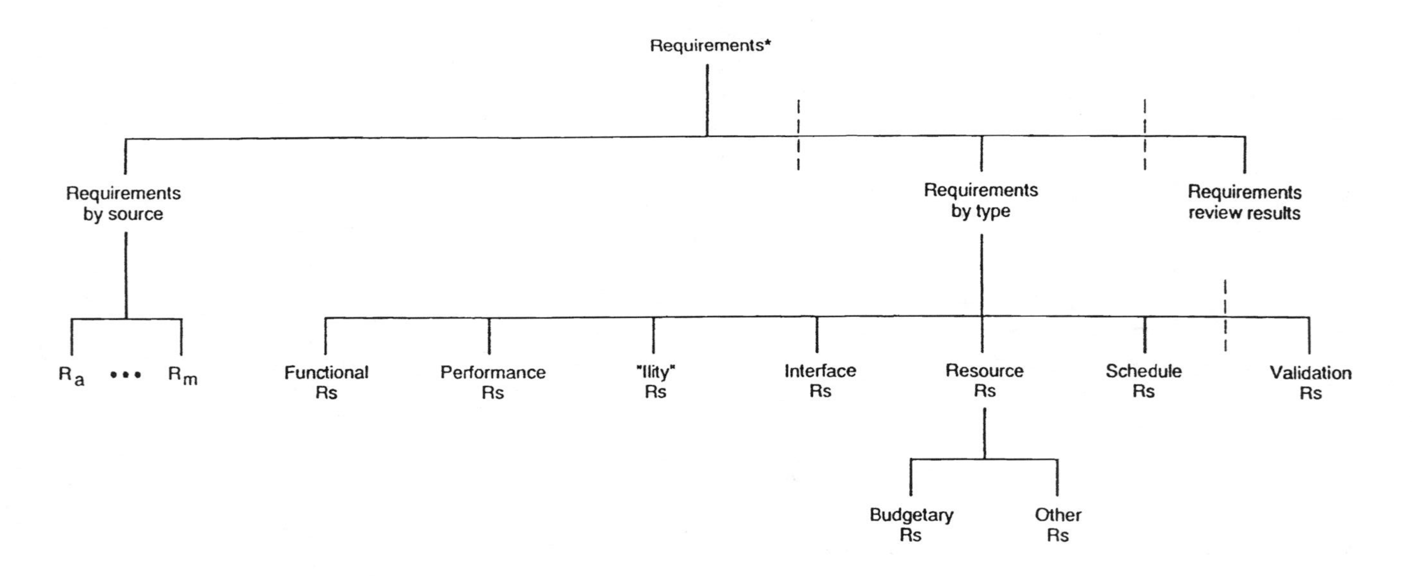

* Requirements are abbreviated as "Rs"

Figure A.3c. The requirements portion of a system development project's work breakdown structure.

view results, which is a corollary to *requirements by type*. As above, short dashed vertical lines indicate chronological relationships:

a. *Requirements by source* is the left-most element at the first level of decomposition and is thus the first element at this level to be prepared.

   Preparing *requirements by source* implies compiling a list of the sources. The scope of work (and contract, if any), applicable regulations, and organizational policies and procedures are basic sources, but these may not be sufficient. In many cases, interviews must also be conducted with customers, regulators, and management to identify all the requirements.

   If a requirement is discovered belatedly, much early work could be rendered useless and wasted. Therefore, as many potential requirements sources as possible should be identified at the beginning. The person responsible for developing the requirements should list the sources that will be consulted, have the list critiqued and, if necessary, correct it. Critiques should be solicited from diverse individuals in order to minimize the chance of overlooking a source.

   Once the list of requirements sources has been developed, the sources can be consulted or interviewed and their requirements collected and tabulated. As part of the interview each source can be asked also to review the list of sources and suggest any additions. Again, the aim is to identify all relevant sources early.

   Despite attempts to obtain all the requirements early, some may not be recognized until later in the project. When it is known at the outset that the initial round of interviews will fall seriously short in defining the ultimate requirements, the project can be organized as a series of incremental or phased deliveries, with each successive delivery producing new information about requirements to be used in the next delivery. In the absence of this approach, changes in requirements must be handled as change requests.

b. *Requirements by type* is the second element at the first level of decomposition to be prepared.

After all the sources have been consulted, the requirements should be collated, categorized, and ranked. If inconsistencies or incompatibilities appear among the requirements, they must be reconciled via negotiations. Upon successful negotiations, the governing or controlling value for each type of requirement can be identified. These latter values can then be recategorized into *requirements by type*. Figure A.3c shows a typical set of types.

Since the project will need to prove at the end that it has satisfied the customer's requirements, it needs validation (or acceptance) requirements. If the requirements determined above can be completely tested directly, then they can serve as their own validation requirements. If, however, direct testing would destroy the product, the number of test combinations and permutations needed for exhaustive testing is too large to be practical, or direct testing is too expensive or time-consuming or just plain impossible, then surrogate requirements based on sampling, models, simulation, analysis, inspection, demonstrations, and so forth are needed as validation requirements.

Because validation requirements become the basis for concluding whether or not the project results satisfy the customer, they must be defined carefully and confirmed with the customer. The time to do this is near the beginning of the project, when the direct requirements are confirmed. When validation requirements are agreed upon early, the project team can then confidently organize its work to satisfy the validation requirements.

c. *Requirements review results* is the third element at the first level of decomposition to be prepared.

The requirements are ready for review when the *requirements by type* reflect all the *requirements by source* and are mutually compatible. A successful review (see Appendix E) means that the *requirements by type,* including the validation requirements, are satisfactory, at least for the time being. As the rest of the project unfolds, the requirements may have to be revised to keep the entire project in harmony.

3. The Design

Figure A.3d is a WBS of the design. This figure can be best described after briefly describing designs and the design process.

A design has five types of information:

a. A list of the subsystems that will be integrated to form the system
b. Statements of each subsystem's requirements
c. Statements of the interfaces between the subsystems
d. A description of how the subsystems are to be assembled or integrated
e. A statement of the verification requirements that will be used to confirm that the subsystems have been assembled or integrated correctly

To arrive at this information, members of the project team need to consider alternatives or options that appear capable of meeting the project's requirements. Options may include *make versus buy* and *a single delivery versus incremental or phased deliveries* as well as technical alternatives. Incremental deliveries are particularly useful when customers are unable to articulate requirements without seeing a physical example.

Of the options that indeed meet the requirements, one must be selected as preferred, i.e., superior to or at least as good as all others. Selecting a preferred option requires both

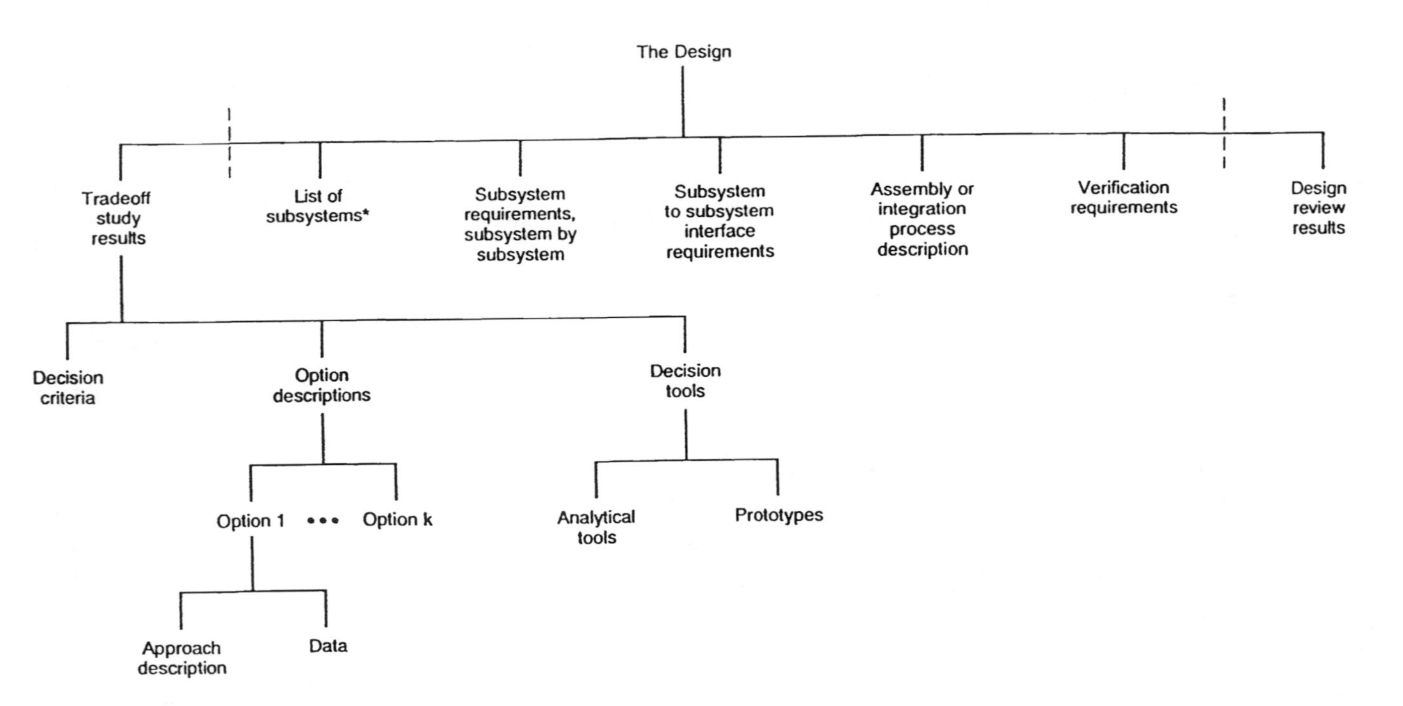

* Subsystems may include system assembly or integration tooling, test equipment, etc.

Figure A.3d. The design portion of a system development project's work breakdown structure.

partitioning guidelines, described in Chapter 3, and weighted decision criteria that have been confirmed with the customer. It also requires decision tools, e.g., models of performance, schedule, and cost; simulations; prototypes; and so forth.

To compare options is to perform tradeoff studies, and the results of this work are shown as first-stage products in the design's WBS, Figure A.3d. This part of the WBS shows the ingredients mentioned in the description above, namely *decision criteria, option descriptions,* and *decision tools.* If all of these ingredients are present, tradeoff studies can be performed to produce *tradeoff study results.*

The products of the second stage are in fact the five types of information listed above. These five items together are the essential characteristics of the preferred or surviving option and they must be mutually compatible. They become *the design* once they have been confirmed by the design review.

The product of the third stage is *design review results.* As above, a successful review (see Appendix E) means that the design is satisfactory, at least for the time being. As the project continues to unfold, the design may have to be adjusted to keep the entire project in harmony.

4. Subsystem i

Figure A.3e is a WBS for a typical subsystem. Each subsystem is identified in the design, as described in the previous section. Comparison of Figures A.3a and A.3e shows that each subsystem has the same array of WBS elements as the system itself. That is, each subsystem can be treated as a system, with its own plan, requirements, design, and so forth. Their WBSs are the same as the corresponding WBSs for the system except that an additional *sub* prefix is added to all systems and subsystems (i.e., *subsystem* instead of *system,* and *sub-subsystem* instead of *subsystem*).

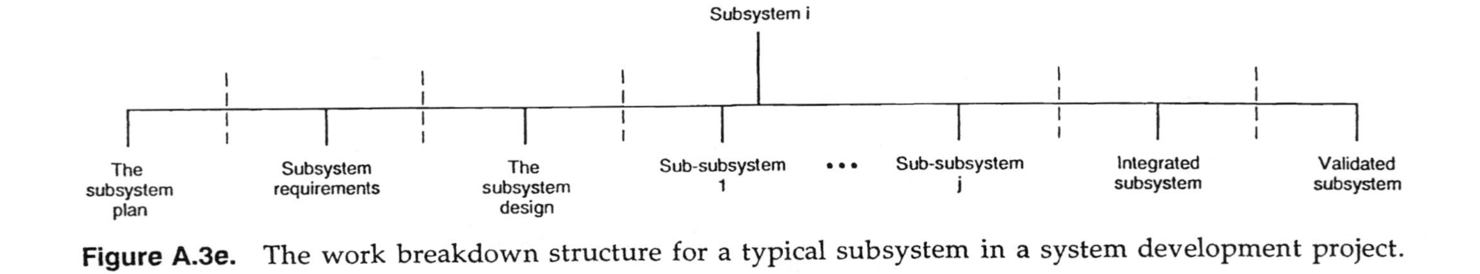

Figure A.3e. The work breakdown structure for a typical subsystem in a system development project.

While everything said about the various parts of the system WBS applies to subsystem WBSs as well, an additional comment is needed about subsystem requirements: Most of a subsystem's requirements are identified in the system design, either as direct requirements or as interface requirements. However, additional subsystem requirements can also appear once the nature of the subsystem is known. For example, if a subsystem can emit radiation in the radio spectrum, then regulations of the Federal Communications Commission (FCC) come into play. If the subsystem can not emit such radiation, then FCC regulations will not apply.

When all the subsystems have passed their individual validation reviews (Appendix E), an implementation phase-end review can be held. A successful review (Appendix E) means that the subsystems are ready *as a set* to be integrated to form the system.

5. The Integrated System and the Validated System

Figures A.3f and A.3g are WBSs for the next-to-last and last parts of the system WBS, respectively. By now, WBSs should speak for themselves and need only a little elaboration.

Both the integrated system and the validated system WBSs incorporate information products from earlier phases. The assembly or integration process description and the verification requirements, which are leading aspects of the integration plan, come from the design. The validation requirements, which are a leading aspect of the validation plan, come from the requirements.

Figures A.3f and A.3g show integration and validation equipment, staff, and so forth as being developed separately from the development of the subsystems, but this need not be the case. An integration plan or validation plan can call for any of its inputs to be a subsystem and thus developed

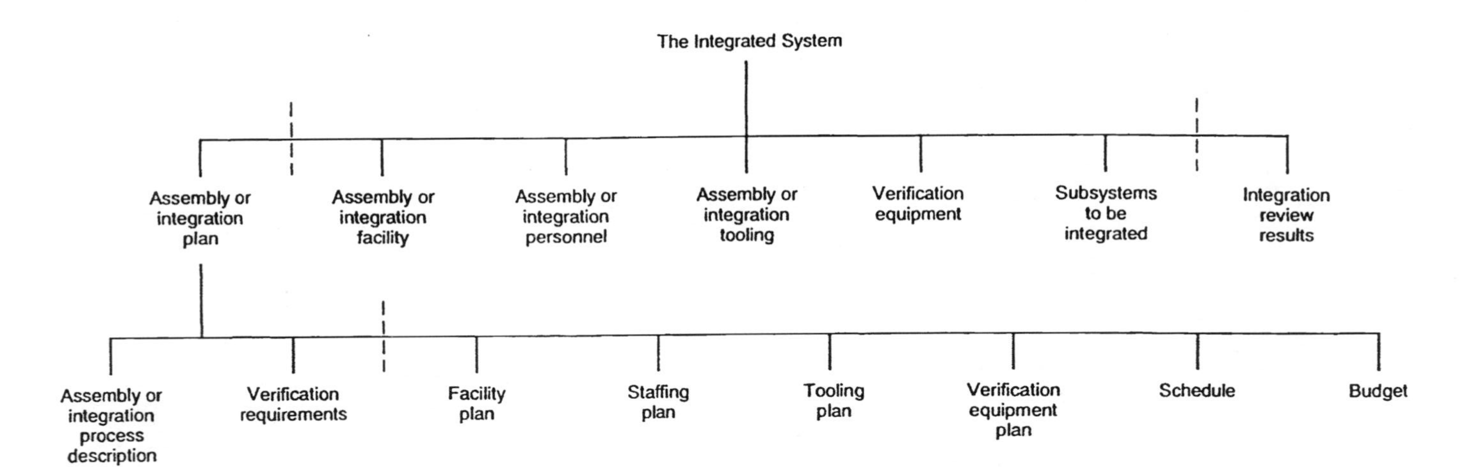

Figure A.3f. The integrated system portion of a system development project's work breakdown structure.

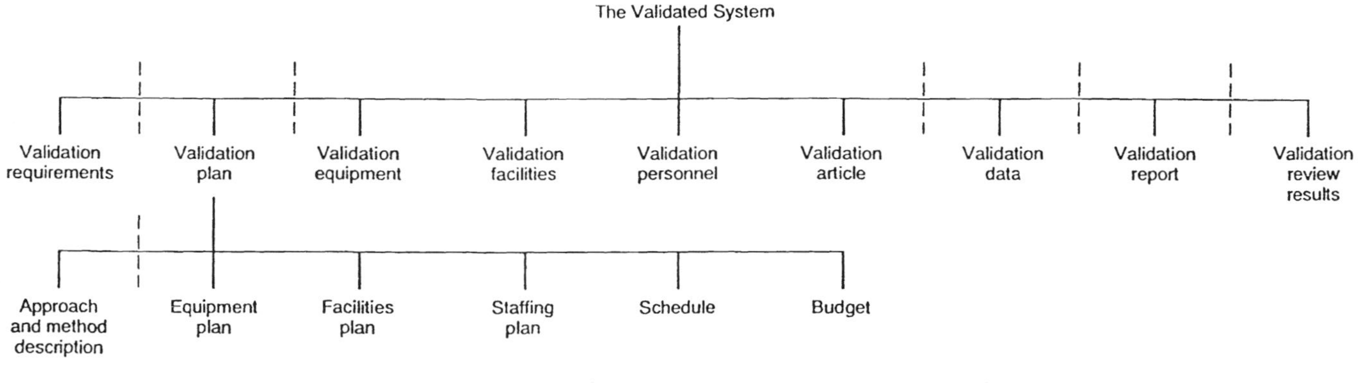

Figure A.3g. The validated system portion of a system development project's work breakdown structure.

in parallel with "real" subsystems. This latter approach enhances coordination of the input's development with development of "real" subsystems and is an option to be evaluated during design tradeoff studies.

Appendix B

Transforming a Work Breakdown Structure into a Precedence Chart

A work breakdown structure (WBS) can be transformed into a precedence chart as follows:

1. Figure B.1a is a WBS. Each item in the figure represents an internal or external product, and lower-level items must be prepared before higher-level items. Also, an item on the right side of the short vertical dashed line cannot be completed until the item at the same level on the left of the dashed line has been completed.
2. Figures B.1b and B.1c are intermediate versions of the precedence chart corresponding to the WBS in Figure B.1a. Each product in the WBS is now represented in the left-to-right arrangement as the activity that creates the product from its subproducts.
3. Figures B.1d, B.1e, and B.1f are alternative final versions of the precedence chart. In all cases, activities to the left precede those to the right.
4. Figures B.1d and B.1e are *activity on arrow* versions of the precedence chart. Figure B.1e is called i-j notation.
5. Figure B.1f is an *activity on node* version of the precedence chart. The nodes are the boxes.

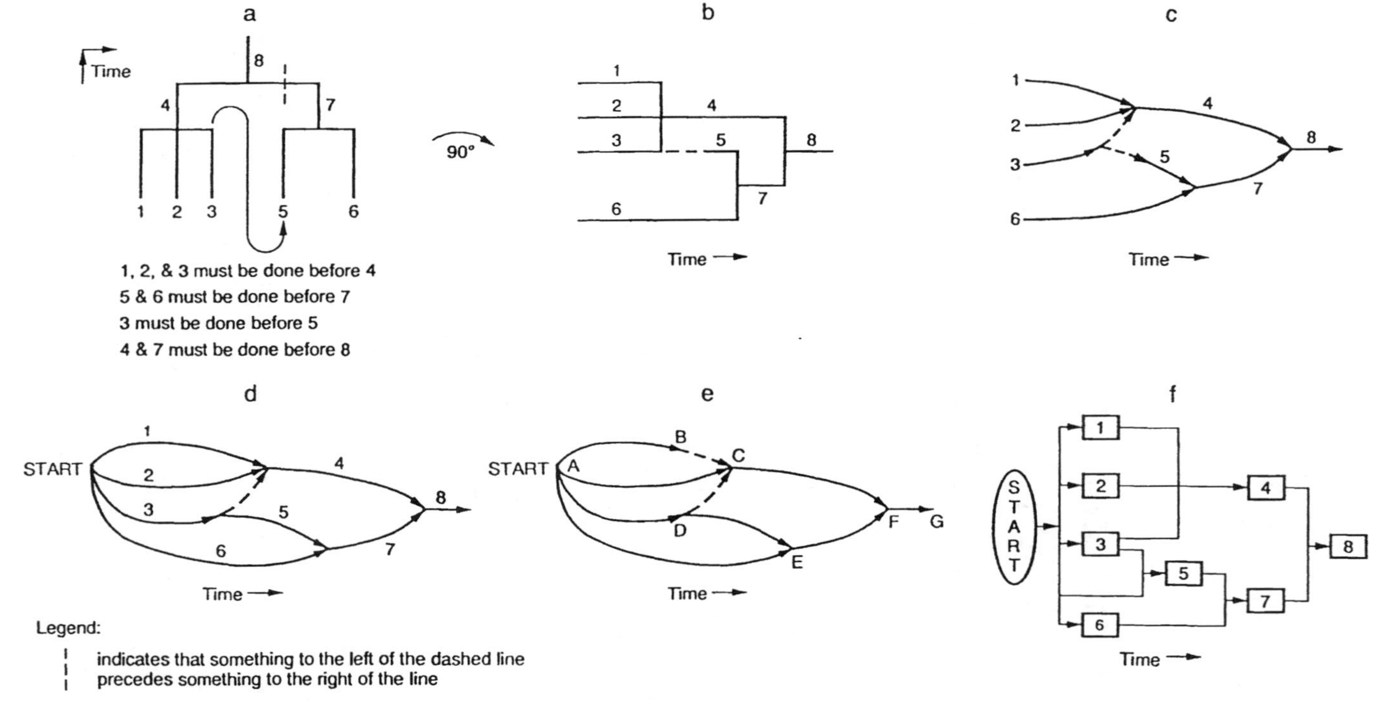

Figure B.1. Transformation of a work breakdown structure into a precedence chart.

The transformation of a WBS into a precedence chart can of course be done by hand, but it need not be. Capable computer-based project management information systems (PMIS) can perform the transformation once the WBS has been created.

Only elements that are in the WBS are incorporated in the precedence chart by the transformation. Other items may have to be inserted in order to make the precedence chart complete. Examples are staffing activities, obtaining authorizations, quality inspections, and so forth.

Appendix C

Cost Versus Time Profiles

The cost versus time profile of most, but not all, projects follows an S-curve, such as the one in Figure C-1. The various parts of the curve are identified by number and have the following characteristics:

Stage 1. Stage 1 is the start-up portion of the project, while the project manager plans, negotiates for staff, and generally sets things up. Costs accumulate slowly during this time unless expensive equipment is purchased.

Stage 2. Stage 2 represents a rapid expenditure of funds as work proceeds at a fast clip. A cost control system that cannot keep up with the expenditures at this time leaves the project manager vulnerable to missed objectives with inadequate resources to recoup.

Stage 3. Stage 3 corresponds to the time when most of the work is complete and attention is devoted to report writing, tying up loose ends, etc. Expenditures are made at a lower rate again.

Stage 4. Stage 4 (or point 4) refers to the time when all explicit commitments have been met. Expenditures to this date should not equal the total budget because there are

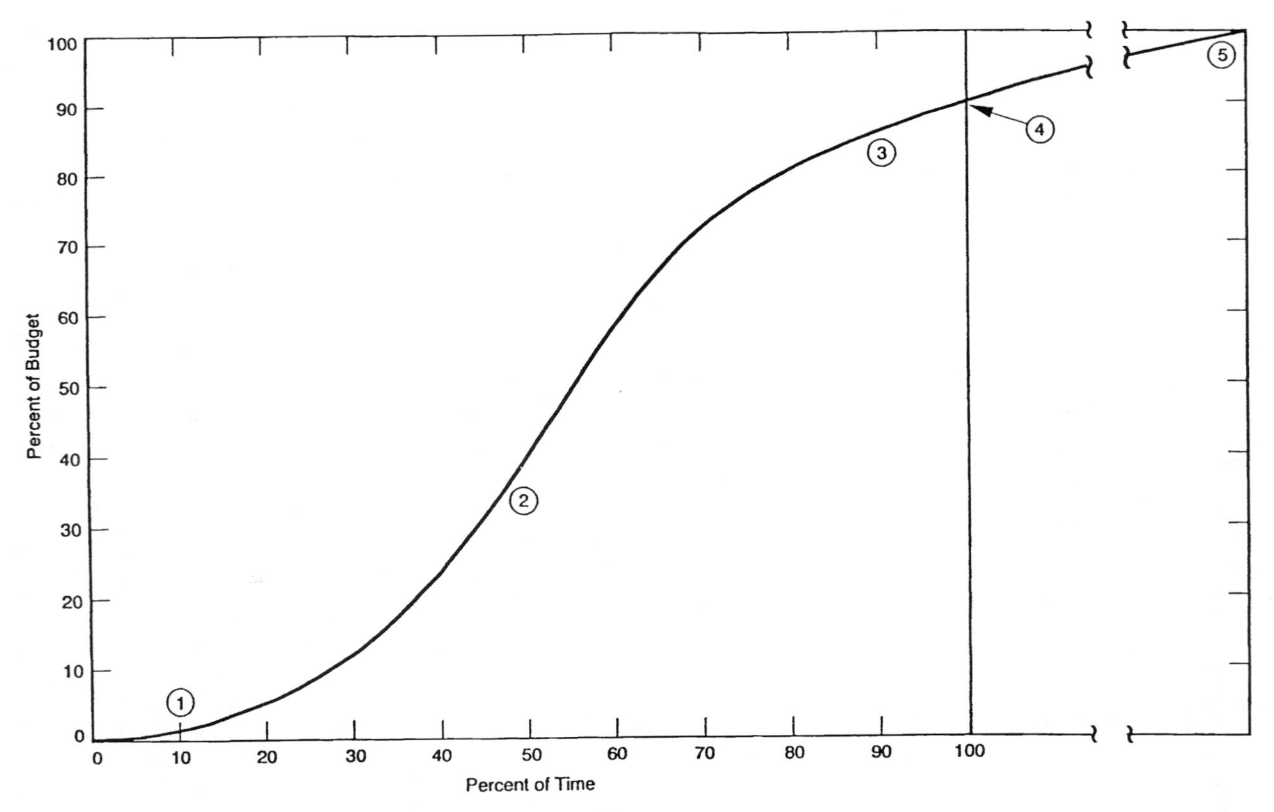

Figure C.1. A typical cost versus time profile

still post-accomplishment activities to be closed out and charged against the project budget.

Stage 5. Stage 5 is the period when the project is finally closed out and minor last costs dribble in. These last costs plus those incurred in accomplishing the project through Stage 4 should not exceed the overall project budget as finally negotiated and approved.

Appendix D

Measuring Project Progress: Percent Complete, Earned Value, and the Size of Work Increments

Two techniques for measuring project progress are *percent complete* and *earned value.* Both techniques require completed work to be measured in increments of appropriate size. This appendix shows how to determine appropriate increments for meaningful *percent complete* and *earned value* measurements.

1. Percent Complete

Percent complete is the ratio of *work done* to [*work done* + *work to be done*] expressed as a percentage. The following steps lead to acceptably small increments for *percent complete* assessments:

a. If we use milestones to refer to substantial pieces of work, we can use inch-pebbles to refer to small pieces or increments of work, with many inch-pebbles needed to make a milestone. The inch-pebbles should all be about the same size, regardless of the milestones to which they pertain.
b. We can use a 0–100 rule for considering an inch-pebble as done. An inch-pebble is counted at 100% if it is com-

pletely done and completely correct. Otherwise, it is counted at 0%.

c. Alternatively, we can use a 50–50 rule for the status of an inch-pebble. An inch-pebble is counted at 100% if it completely done and completely correct. It is counted at 50% if it is started but not completely done correctly. And it is counted at 0% if it has not been started.
d. If we assume inch–pebbles of a given size and have a schedule for their accomplishment, we can determine the project's percent complete as of some randomly chosen date two ways, i.e., using the 0–100 rule and using the 50–50 rule.
e. The two values of percent complete determined in step 1.d can be differenced and they can be averaged. Call their difference A and their average B.
f. Half of A/B indicates the plus and minus error band associated with using the 0–100 rule and the 50–50 rule for inch-pebbles of the assumed size; it can be expressed as a percentage.
g. When the error band determined in step 1.f is acceptably small, the inch-pebble size assumed in step 1.d is acceptably small.
h. The smaller the inch-pebbles, the greater is their number and therefore the greater is the effort needed to measure how many of them are done at any time.
i. By assuming inch-pebbles of various sizes, we can find a size that is small enough (steps 1.d through 1.g) but is not so small that it causes needless measurements (step 1.h).

2. Earned Value

Earned value refers to the *budgeted cost of work peformed* (BCWP), where *work performed* is the same as *work done* above. The term *earned value* comes from using BCWPs as the basis for payments from the customer to the project organization.*

The following steps lead to determining acceptably sized increments:

a. Again, we can use milestones to refer to substantial pieces of work and inch-pebbles to refer to small pieces or increments of work, with many inch-pebbles needed to make a milestone. The inch-pebbles within a billing period should all be about the same size, regardless of the milestones to which they pertain.
b. BCWPs work best when the 0–100 rule (step 1.b) is used. A fractional rule such as the 50–50 rule (step 1.c) is subject to abuse, e.g., by barely starting a lot of inch-pebbles just to collect half of their BCWP.
c. If we assume inch-pebbles of a given size and have a schedule for their accomplishment within a billing period, we can determine two numbers: (1) the number of inch-pebbles that should be finished within the billing period, and (2) the number of inch-pebbles whose start and finish dates straddle the last day of the billing period. Call the first of these numbers C and the second D.
d. Half of D/C indicates the plus and minus error band on BCWP or earned value; it can be expressed as a percentage.

---

*BCWPs are also used to determine if work is proceeding overall as scheduled. This is done by comparing the BCWP for a project or project element with its *budgeted cost of work scheduled* (BCWS). A hazard lies in this approach: Unless critical path work is examined specifically and by itself in a BCWP-BCWS comparison, lagging critical path work can be masked by accomplishments elsewhere that are ahead of schedule. In such cases, a favorable BCWP-BCWS comparison can lead to a false sense of well-being.

Comparing a BCWP for a project or project element with its *actual cost of work performed* (ACWP) is the same as comparing its planned and actual costs, which is discussed in Chapter 4.

e. As the size of the inch-pebbles is decreased, C will increase and D might decrease and the BCWP error band will decrease.
f. When the BCWP error band determined in step 2.d is acceptably small, the size of the inch-pebbles assumed in step 2.c is acceptably small.
g. The smaller the inch-pebbles, the greater is their number and therefore the greater is the effort needed to measure how many of them are actually done at the end of the billing period.
h. By assuming inch-pebbles of various sizes, we can find a size that is small enough (steps 2.c through 2.f) but is not so small that it causes needless measurements (step 2.g).

If both *percent complete* and *earned value* are to be used but do not require the same size inch-pebbles, then the smaller size should be selected for both applications in order to minimize monitoring cost.

Appendix E

Questions to Assess Work Done

The following questions can be used to assess work done on system development projects. They are organized by the phases shown in the work breakdown structure (WBS) described in Appendix A.

1. Planning Phase

   a. Is the management plan workable?
   b. Are the strawman requirements and design defined sufficiently to permit cost and schedule estimation and project planning?
   c. Are the cost and schedule estimates appropriate to the maturity of the requirements and the design?
   d. Is the design compatible with the requirements, budget, and schedule?
   e. Can the WBS and account structure accommodate a suitable range of plausible designs?
   f. Have arrangements been made for the participation of all personnel who are needed to do concurrent engineering?
   g. Has a quality control plan been included in the overall plan?

h. Are risk assessment and risk reduction included in the risk management plan?
i. Do the budget, schedule, and scope (as a set) include suitable contingency allowances?
j. Will the reporting and review plan provide the level of control needed by supersystem?
k. Can the implementation plan accommodate the development as it might unfold?
l. Are staffing, tools, and training plans compatible?
m. Can the documentation plan accommodate a suitable range of plausible development arrangements?
n. Are subplans, e.g., the plan for obtaining permits, the resource expenditure profile, the quality assurance plan, etc., compatible with each other and consistent with the overall plan?

2. Requirements Phase

a. Are assumptions previously made still valid?
b. Were *all* personnel needed for concurrent engineering involved in defining the functional, performance, "-ility," interface, and validation (FPIIV) requirements?
c. Were an operations concept and a logistics scenario used in defining the requirements?
d. Are *testability* requirements included in the requirements?
e. Are the FPIIV requirements well-defined, complete, compatible with each other, and agreed to, both within the project and with the customer?
f. Is the strawman design defined sufficiently to permit cost and schedule estimation?
g. Are the requirements compatible with personnel resources, budget, and schedule?
h. Have a quality control approach and plan been defined, found compatible with the project budget and schedule, and agreed to?

   i. Does the WBS and account structure still accommodate the range of plausible designs?

3. Design Phase

   a. Are assumptions previously made still valid?
   b. Have design tradeoff studies been performed?
   c. Does the design satisfy *testability* requirements?
   d. If applicable, is the design operable, maintainable, and economical in service?
   e. Is the design well defined and agreed to?
   f. Does the design respond to *all* customer requirements?
   g. Is the design compatible with available resources and schedule?
   h. Is the design robust, i.e., can it easily accommodate moderate changes in requirements, resources, and schedule?
   i. Have compatible sets of subsystem requirements, budgets, and schedules been negotiated?
   j. Have subsystem-to-subsystem interfaces been defined?
   k. Has the process for integrating subsystems to form the system been described?
   l. Have system verification requirements been defined?
   m. Do the requirements appear to be stable?
   n. Are the WBS and account structure still appropriate in light of the system design?

4. Implementation Phase

   a. Are assumptions previously made still valid?
   b. Will the system as implemented be operable, maintainable, and economical in service?
   c. Can the system be validated?
   d. Has a suitable validation plan been prepared?
   e. Are satisfactory arrangements in place to integrate the system?

f. Have all the subsystems been validated individually?
g. Are the requirements and the design still valid?
h. Are change orders being closed out at a rate that equals or exceeds their initiation?
i. Are the WBS and account structure, staffing plan, budget, and schedule still appropriate in light of subsystem developments?
j. Are all subsystem managers meeting their commitments?

5. Integration Phase

a. Are assumptions previously made still valid?
b. Has the system been built as designed?
c. Does the system show promise of working as required?
d. Will operations and maintenance costs be as expected?
e. Are arrangements and documentation in place to validate the system?
f. Is the problem-reporting mechanism working properly?
g. Are there any adverse trends?
h. Are subsystem personnel adequately supporting integration of the system?
i. Have all problems discovered during integration been resolved?

6. Validation Phase

a. Are assumptions previously made still valid?
b. Is the problem-reporting mechanism working properly?
c. Have all problems discovered during validation been resolved?
d. Are subsystem personnel adequately supporting validation of the system?
e. Are there any adverse trends?

f. Is all documentation needed for commissioning and operations in hand?
g. Is the system ready to be commissioned?

7. Commissioning Phase (not included in the WBS in Appendix A)

   a. Was the commissioning successful?
   b. Has the operating staff been trained and certified?
   c. Have spares and other logistical items been provided?
   d. Have all problems discovered during commissioning been resolved?
   e. Has an action plan been prepared for resolving residual problems?
   f. Is the customer ready to accept responsibility for the system?

Appendix F

Cultural Aspects of Successful Matrix Organizations

The culture in a successful matrix organization supports advanced planning, open communication, collaboration and cooperation rather than competition, resource-sharing, win-win negotiation, timely conflict resolution, and problem-solving rather than formality. Establishing and maintaining this culture is chiefly the responsibility of the organization's highest boss, who sets the tone by personal example and by prescribing and enforcing appropriate standards.

In a successful matrix organization, the mechanisms and behaviors that are described in Chapter 7 receive more than lip service. For example, resources or time may seem too scarce both to serve all immediate customers immediately and also to plan, negotiate conflicts, and behave according to other matrix norms. Yet, adequate planning, negotiation, etc., must still be done. Otherwise, resources will likely not be arranged well enough for efficient use and the scarcity will be perpetuated. In this event, matrix success will be frustrated and management's commitment to the matrix will be doubted. Such doubts can lead to poor matrix behavior even if resource competition should lessen.

A suitable culture is fostered by suitable personal behavior, and the reverse is also true. This means that desired behaviors are encouraged and rewarded and counterproductive behaviors are disciplined. This may be easier said than done. For example, if some multiple bosses have personally thrived in a competitive atmosphere and are uncomfortable relying on cooperation, then they may feel that their own success will be compromised by cooperating instead of competing. Competitive behavior, however, will frustrate overall success and should be discouraged, while cooperative behavior must be encouraged. To encourage cooperation and discourage internal competition, higher bosses must adjust reward systems conspicuously. This means of course that the higher bosses must themselves believe in cooperation, which might not occur if they consider internal competition essential to organizational success.

Achieving a culture that supports matrix arrangements must start at the top of the organization. If the highest bosses do not wholeheartedly accept the conditions necessary for effective matrix arrangements and actively promote them throughout the organization, then matrix success will be stymied. Nothing else can substitute for their critical behavior.

Bibliography

Archibald, R. D., *Managing High-Technology Programs and Projects,* 2nd Edition, New York: John Wiley and Sons, Inc. (1992), xv + 384 pp.

Ashley, D. B., Urie, C. S., and Jaselskis, E. J., "Determinants of Construction Project Success," *Project Management Journal,* 18, No. 2 (June 1987), 69–79.

Baker, S., and Baker, K., *On Time/On Budget,* Englewood Cliffs, NJ: Prentice-Hall, Inc. (1992), xv + 304 pp.

Bergantz, J. L., "Alternative Contracting/Acquisition Strategies within DOD," *Program Manager,* 18 (September/October 1989), 32–36.

Blanchard, F. L., *Engineering Project Management,* New York and Basel: Marcel Dekker, Inc. (1990), ix + 246 pp.

Casse, P., *Training for the Cross-Cultural Mind,* 2nd Edition, Yarmouth, ME: Intercultural Press (1981), xxvi + 260 pp.

Casse, P., *Training for the Multicultural Manager,* Yarmouth, ME: Intercultural Press (1982), xv + 191 pp.

Chambers, G. J., "The Individual in a Matrix Organization," *Project Management Journal,* 20, No. 4 (December 1989), 37–42, 50.

Cleland, D. I., *Project Management: Strategic Design and Implementation,* 2nd Edition, New York: McGraw-Hill Book Company (1994), xx + 478 pp.

Cleland, D. I., and King, W. R., Editors, *Project Management Handbook*, 2nd Edition, New York: Van Nostrand Reinhold Company (1988), x + 997 pp.

Davis, S. M., and Lawrence, P. R., *Matrix*, Reading, MA: Addison-Wesley Publishing Co., Inc. (1977), xvii + 235 pp.

Department of Defense, *Defense Acquisition Management Policies and Procedures*, Instruction No. 5000.2, Springfield, VA: National Technical Information Service (1991).

Dimensions of Project Management: Fundamentals, Techniques, Organization, Applications, Berlin and New York: Springer-Verlag (1990), xvii + 336 pp.

Doyle, M., and Straus, D., *How to Make Meetings Work: The New Interaction Method*, New York: Playboy Paperbacks, Div. of P.E.I. Books, Inc. (1977), x + 301 pp.

Dunsing, R. J., *You and I Have Simply Got to Stop Meeting This Way*, New York: American Management Association (1977), 88 pp.

Fisher, R., and Ury, W., *Getting to Yes: Negotiating Agreement Without Giving In*, Boston: Houghton Mifflin Co. (1981), xiii + 163 pp.

Fleming, Q. W., *Cost/Schedule Control Systems Criteria*, Revised Edition, Chicago and Cambridge (England): Probus Publishing Company (1992), xii + 564 pp.

Frankel, E. G., *Project Management in Engineering Services and Development*, London and Boston: Butterworths & Co., Ltd. (1990), x + 358 pp.

Gadeken, O. C., "The Right Stuff: Results of DSMC's Program Manager Competency Study," *Program Manager*, 18 (September/October 1989), 22–25.

Gobeli, D. H., and Larson, E. W., "Project Management Problems," *Engineering Management Journal*, 2, No. 2 (June 1990), 31–36.

Guiterrez, G. J., and Kouvelis, P., "Parkinson's Law and Its Implications for Project Managers," *Management Science*, 37, No. 8 (August 1991), 990–1001.

Hamburger, D. H., "On-Time Project Completion—Managing the Critical Path," *Project Management Journal*, 18, No. 4 (September 1987), 79–85.

Harrison, F. L., *Advanced Project Management,* 2nd Edition, New York: Halstead Press (1985), 374 pp.

Johnson, R. K., "Program Control from the Bottom Up—Exploring the Working Side," *Project Management Journal,* 16 (March 1985), 80–88.

Kerzner, H., *Project Management: A Systems Approach to Planning, Scheduling, and Controlling,* 4th Edition, New York: Van Nostrand Reinhold Company (1992), xv + 1023 pp.

Kezbom, D. S., Schilling, D. L., and Edward, K. A., *Dynamic Project Management: A Practical Guide for Managers and Engineers,* New York: John Wiley and Sons, Inc. (1989), ix + 357 pp.

Kimmons, R. L., *Project Management Basics,* New York and Basel: Marcel Dekker, Inc. (1990), v + 338 pp.

Kimmons, R. L., and Loweree, J. H., Editors, *Project Management: A Reference for Professionals,* New York and Basel: Marcel Dekker, Inc. (1989), xvii + 1096 pp.

Klimstra, P. D., and Potts, J., "What We've Learned Managing R&D Projects," *Research Technology Management,* 31, No. 3 (May/June 1988), 23–39.

Kloppenborg, T., and Mantel, S. J., Jr., "Tradeoffs on Projects: They May Not Be What You Think," *Project Management Journal,* 21, No. 1 (March 1989), 13–20.

Knutson, J., and Bitz, I., *Project Management,* New York: American Management Association (1991), ix + 198 pp.

Lake, J., "The Work Breakdown Structure: It's Much More Than a Cost Reporting Structure," *Program Manager* (July/August 1993), 3–9.

Levine, H. A., "Look for Differences in Project Management Software!" *PM Network,* 6 (October 1992), 27–30.

Lewis, W. M., and Jens, R. M., "Project Management Lessons from the Past Decade of Mega-Projects," *Project Management Journal,* 18, No. 5 (December 1987), 69–74.

Miller, S., Wackman, D., Nunnally, E., and Saline, C., *Straight Talk,* New York: Rawson, Wade Publishers, Inc. (1981), 340 pp.

Mustafa, M. A., and Murphee, E. L., "A Multicriteria Decision Support Approach for Project Compression," *Project Management Journal,* 20, No. 2 (June 1989), 29–34.

Myers, I. B., *Introduction To Type,* 5th Edition, Palo Alto, CA: Consulting Psychologists Press, Inc. (1993), ii + 30 pp.

Myers, I. B., and Myers, P. B., *Gifts Differing,* Palo Alto, CA: Consulting Psychologists Press, Inc. (1990), xiii + 217 pp.

National Aeronautics and Space Administration, *Management of Major System Programs and Projects,* NASA Handbook 7120.5, Washington, DC: National Aeronautics and Space Administration (November 1993).

Nicholas, J. M., *Managing Business and Engineering Projects: Concepts and Implementation,* Englewood Cliffs, NJ: Prentice-Hall, Inc. (1990), xv + 543 pp.

Nierenberg, G. I., *The Art of Negotiating,* New York: Cornerstone Library, Inc. (1980), 192 pp.

Oncken, W., "The Authority to Manage," *Manage,* 18, No. 8 (June/July 1966), 4–13.

Phillips, M. E., Goodman, R. A., and Sackmann, S. A., "Exploring the Complex Cultural Milieu of Project Teams," *PM Network,* 6, (November 1992), 20–22, 24–26.

Pinto, J. K., and Slevin, D. P., "The Project Champion: Key to Implementation Success," *Project Management Journal,* 20, No. 4 (December 1989), 15–20.

Pitts, C. E., "For Project Managers: An Inquiry into the Delicate Art and Science of Influencing Others," *Project Management Journal,* 21, No. 1 (March 1989), 21–24, 42.

"PM 101" The Project Manager-A Leader," *PM Network,* 7 (December 1993), 28–31.

Porter, E. H., "Strength Deployment Inventory Manual of Administration and Interpretation," Pacific Palisades, CA: Personal Strengths Publishing, Inc. (1985), 22 pp.

Quality in the Constructed Project, Manual of Professional Practice No. 73, New York: American Society of Civil Engineers (1990), 144 pp.

Randolph, W. A., and Posner, B. Z., *Effective Project Planning and Management,* Englewood Cliffs, NJ: Prentice-Hall, Inc. (1988), x + 163 pp.

Reiss, G., *Project Management Demystified: Today's Tools and Techniques,* London and New York: Chapman and Hall, Ltd. (1992), viii + 213 pp.

Ritz, G. J., *Total Engineering Project Management,* New York: McGraw-Hill Publishing Co. (1990), xiv + 370 pp.

Ruskin, A. M., "A Further Note on Monitoring and Contingency Allowances: Complementary Aspects of Project Control," *Project Management Journal*, 16, (December 1985), 52–56.

Ruskin, A. M., "Project Management and System Engineering: A Marriage of Convenience," *PM Network*, 5, No. 5 (July 1991), 38–41.

Ruskin, A. M., "Twenty Questions That Could Save Your Project," *IEEE Transactions on Engineering Management*, EM-2, No. 1 (January 1986) 3–9.

Ruskin, A. M., and Estes, W. E., "The Project Management Audit: Its Role and Conduct," *Project Management Journal*, 16 (August 1985, Special Summer Issue), 64–70.

Ruskin, A. M., and Estes, W. E., "Project Risk Management," *PM Network*, 6, No. 3 (April 1992), 30, 33–37.

Rutherford, R. D., *The 25 Most Common Mistakes Made in Negotiating and What You Can Do About Them*, Boulder, CO: Keneric Publishing Co. (1986), ii + 109 pp.

Samid, G., *Computer-Organized Cost Engineering*, New York and Basel: Marcel Dekker, Inc. (1990), viii + 424 pp.

Slemaker, C . M., *The Principles and Practice of Cost/Schedule Control Systems*, Princeton, NJ: Petrocelli Books (1985), xii + 427 pp.

Tatum, C. B., "Barriers to the Rational Design of Project Organizations," *Project Management Journal*, 15 (December 1984), 53–61.

Teplitz, C. J., and Amor, J-P, "Improving COM's Accuracy Using Learning Curves," *Project Management Journal*, 24, No. 4 (December 1993), 15–19.

Thamhain, H. J., *Engineering Program Management*, New York: John Wiley and Sons, Inc. (1984), xii + 351 pp.

Thamhain, H. J., "Validating Technical Project Plans," *Project Management Journal*, 20, No. 4 (December 1989), 43–50.

Urquhart, G. A., "A Project Manager's Exposure When Claims Arise on a Construction Project," *Project Management Journal*, 15, (September 1984), 36–38.

Ury, W., *Getting Past No*, New York: Bantam Books (1991), 161 pp.

Westney, R. E., *Managing the Engineering and Construction of Small Projects*, New York and Basel: Marcel Dekker, Inc. (1985), xi + 283 pp.

Index

About the Authors

ARNOLD M. RUSKIN is a founding partner of Claremont Consulting Group, La Cañada, California, and provides consulting and training in project management, system engineering, and engineering organization and management. He has managed system, process, hardware, and software design and development projects and environmental impact assessments and has been director of the Engineering Executive Program at UCLA and professor of engineering at Harvey Mudd College. Dr. Ruskin is the author of over 35 professional papers and serves on the editorial review board of *Project Management Journal* and the editorial board of *Engineering Management Journal.* An Associate Fellow of the American Institute of Aeronautics and Astronautics and a member of the Project Management Institute, the American Society for Engineering Management, and the National Council on Systems Engineering, Dr. Ruskin is a Registered Professional Engineer, a certified Project Management Professional, and a Certified Management Consultant. He received the Ph.D. degree in engineering materials from the University of Michigan, Ann Arbor.

W. EUGENE ESTES is a Consulting Civil Engineer in Westlake Village, California, and has been a partner and Director of Project Management at Dames and Moore, Los Angeles, California. His professional career has spanned over 40 years and been largely concerned with the direct management of projects, the supervision of project managers, and the auditing of projects. Work performed under his supervision includes research, development, and design projects; geotechnical investigations; major maintenance efforts; and the entire spectrum of construction programs, from feasibility studies through start-up of operations. Mr. Estes is a Life Member of the American Society of Civil Engineers and was a member of the Project Management Institute and the Society of American Military Engineers. He is a Registered Professional Engineer and received the M.S. degree in civil engineering from the University of Illinois, Urbana.